Book 2

Embryonic and Fetal Development

D0144871

REPRODUCTION IN MAMMALS

Book 2

Embryonic and Fetal Development

EDITED BY

C. R. AUSTIN

Fellow of Fitzwilliam College,
Charles Darwin Professor of Animal Embryology,
University of Cambridge

AND

R. V. SHORT

Fellow of Magdalene College,
Reader in Reproductive Biology,
University of Cambridge

ILLUSTRATIONS BY JOHN R. FULLER

CAMBRIDGE UNIVERSITY PRESS

Published by the Syndics of the Cambridge University Press
Bentley House, 200 Euston Road, London NW1 2DB
American Branch: 32 East 75th Street, New York, N.Y. 10022

© Cambridge University Press 1972

ISBN: 0 521 08373 7   hard covers
      0 521 09682 0   paperback

First published 1972
Reprinted 1973

First printed by offset in Great Britain by
Alden & Mowbray Ltd
at the Alden Press, Oxford

Reprinted in Malta by St Paul's Press Ltd

Contents

Contributors to Book 2

C. R. Austin,
Physiological Laboratory,
Downing Street,
Cambridge.

R. L. Gardner,
Physiological Laboratory,
Downing Street,
Cambridge.

G. C. Liggins,
National Women's Hospital,
Claude Road,
Auckland,
New Zealand.

Anne McLaren,
Institute of Animal Genetics,
West Mains Road,
Edinburgh.

R. V. Short,
Department of Veterinary Clinical Studies,
Madingley Road,
Cambridge.

Preface

Reproduction in Mammals is intended to meet the needs of under-graduates reading Zoology, Biology, Physiology, Medicine, Veterinary Science and Agriculture, and as a source of information for advanced students and research workers. It is published as a series of five small text books dealing with all major aspects of mammalian reproduction. Each of the component books is designed to cover independently fairly distinct subdivisions of the subject, so that readers can select texts relevant to their particular interests and needs, if reluctant to purchase the whole work. The contents lists of all the books are set out on the next page.

The present volume on *Embryonic and Fetal Development* treats the progress of the new individual from the zygote stage, through cleavage, differentiation, implantation, placenta formation and fetal growth, to the events immediately preceding and succeeding birth. In addition, a chapter is devoted to the manipulative approach to problems of embryology, which has produced highly significant results in recent years, and another sets out briefly the nature and consequences of anomalous development.

Books in this series

Book 1. *Germ Cells and Fertilization*
Primordial germ cells.   T. G. Baker
Oogenesis and ovulation.   T. G. Baker
Spermatogenesis and the spermatozoa.   V. Monesi
Cycles and seasons.   R. M. F. S. Sadleir
Fertilization.   C. R. Austin

Book 2. *Embryonic and Fetal Development*
The embryo.   Anne McLaren
Sex determination and differentiation.   R. V. Short
The fetus and birth.   G. C. Liggins
Manipulation of development.   R. L. Gardner
Pregnancy losses and birth defects.   C. R. Austin

Book 3. *Hormones in Reproduction*
Reproductive hormones.   D. T. Baird
The hypothalamus.   B. A. Cross
Role of hormones in sex cycles.   R. V. Short
Role of hormones in pregnancy.   R. B. Heap
Lactation and its hormonal control.   A. T. Cowie

Book 4. *Reproductive Patterns*
Species differences.   R. V. Short
Behavioural patterns.   J. Herbert
Environmental effects.   R. M. F. S. Sadleir
Immunological influences.   R. G. Edwards
Aging and reproduction.   C. E. Adams

Book 5. *Artificial Control of Reproduction*
Increasing reproductive potential in farm animals.   C. Polge
Limiting human reproductive potential.   D. M. Potts
Chemical methods of male contraception.   Harold Jackson
Control of human development.   R. G. Edwards
Reproduction and human society.   R. V. Short
The ethics of manipulating reproduction in man.   C. R. Austin

1 The embryo
Anne McLaren

The mammalian egg, though small compared with the eggs of other vertebrates (e.g. the ostrich), is one of the largest cells in the mother's body. Before fertilization, it is also one of the most sluggish cells in the body, with a metabolic rate scarcely higher than a very inert tissue such as bone. It possesses individuality, in the sense that its complement of genes constitutes a unique selection from the maternal set, but its genetic information has not yet been used and hence in all inherited characters it resembles other maternal cells.

Three to 4 days after fertilization of the egg, the embryo may still contain less than a hundred cells, but these cells now approximate to those of an adult organism in average size. The pattern of RNA synthesis and the capacity for synthesizing protein are also similar to those of the adult, while the metabolic rate is as high as the most metabolically active tissues of the maternal body such as the retina. The embryo's own genes have begun to function, coding for proteins which may potentially be recognized by the mother as foreign. Differentiation has begun, giving rise to a tissue, the trophoblast, highly specialized for interaction with the mother.

These striking changes form the subject matter of the first part of this chapter. Since much of what we know about mammalian embryos comes from experiments in which they have been cultivated outside the mother's body, the process of *in vitro* cultivation will also be described, together with the possibilities it offers for experimental analysis and manipulation. The last part of the chapter deals with implantation. The embryo, after a period of relative independence, must become intimately united with the maternal organism if pregnancy is to continue; but by now mother and embryo are two genetically distinct individuals, so that the process of implantation is both complex and hazardous.

The diameter of the egg is about 0.1 mm in all mammalian species, just at the limit of visibility to the unaided eye. It is surrounded by a thin non-cellular envelope, the zona pellucida. The 1-cell stage, because of its relatively large size, is distinguished by a low ratio of nuclear to cytoplasmic material. This ratio, critical for the genetic control of the cell at later stages, is restored to a value resembling that of adult cells by the process of cleavage, in which several successive cell divisions occur without any increase in total mass. Indeed, growth may be 'negative' during cleavage: the total amount of cellular material decreases by about 20 per cent in the cow, and by as much as 40 per cent in the sheep, while the protein content of the mouse embryo falls by 25 per cent during the first 3 days after fertilization.

Cleavage takes place much more slowly in mammals than in most lower vertebrates or invertebrates. Frog eggs cleave about once an hour, and goldfish eggs every 20 minutes or so; but a mouse egg takes 24 hours for its first cleavage division, and 10–12 hours for each succeeding division. Other mammalian embryos that have been examined show the same slow tempo. We do not know the reason for this difference.

Variation in rate of cleavage is common, both among embryos and among cells (blastomeres) of a single embryo. One consequence is that the initial synchrony of the cleaving embryo soon vanishes. Two- and 4-cell stages are much more often encountered than 3- and 5-cell stages; the following day 8-cell stages predominate, but the scatter is wider; after four or five successive cleavage divisions, little synchronization remains. By this stage also, the cells of the outer layer are dividing more slowly than those in the middle.

During the first few cleavage divisions, each mitosis is

followed immediately by DNA synthesis in the two daughter cells, i.e. the pause (G1 period) before DNA synthesis, characteristic of adult cells, is absent. Recent studies on mouse embryos have shown that no G1 period can be detected before the 8- to 16-cell stage.

BLASTOCYST FORMATION

Within the solid ball of cells, the morula, formed by successive cleavage divisions, a fluid-filled cavity appears. The cavity (blastocoele) enlarges rapidly, until the embryo resembles a hollow sphere, the blastocyst, with a single peripheral layer of large flattened cells, the trophoblast layer, and a knob of smaller cells to one side of the central cavity. This knob, the so-called inner cell mass, will give rise to the adult organism, while the cells of the trophoblast form the placenta and embryonic membranes.

Up to this stage, the embryos of all mammals resemble one another closely. Blastocysts, however, develop rather differently in different groups. In the mouse, and probably also in man, the blastocyst cavity begins to form when no more than 20–30 cells are present, and at the time that implantation begins the blastocyst contains no more than 100 or so cells. Kangaroos are similar. In rabbits, blastocyst formation does not begin until about three cleavage divisions later than in the mouse, while the mature blastocyst contains several thousand cells and measures 3–4 mm in diameter. In sheep and cattle, the blastocyst gradually elongates, and may attain a length of 20 cm before attachment in the 2nd or 3rd week of pregnancy. In pigs, this process is accentuated: between the 9th and the 16th day of pregnancy the blastocyst undergoes a 300-fold elongation, changing from a small spherical vesicle to an exceedingly elongated thread-like tube more than a metre in length, before attachment begins.

The time at which the blastocyst cavity starts to form is not related to the size of the embryo. If three blastomeres in a 4-cell

3

mouse or rabbit embryo are destroyed, the survivor continues to develop; the blastocyst cavity will form at the usual time, giving rise to a blastocyst a quarter the size of a normal one. If an embryo is disaggregated into single cells, a fluid-filled vesicle often appears in individual cells at the time that the blastocoele would normally form, suggesting that the fluid originates by intracellular accumulation. The timing of fluid secretion may be related to the number of cleavage divisions that have elapsed since fertilization, in other words to the nucleo-cytoplasmic ratio. This is suggested by the fact that haploid mouse embryos which result from parthenogenetic activation of the unfertilized egg and which have a nucleo-cytoplasmic ratio of half the normal value at any particular cleavage stage, show a delay of about one cleavage division in the formation of the blastocyst cavity. (Parthenogenetic activation is discussed in Chapter 4 of Book 1.)

The most striking feature of the blastocyst is its differentiation into trophoblast and inner cell mass. Differentiation is the central unsolved mystery of development, that from a single fertilized egg develop all the tissues of the adult organism, including such diverse types as bone and brain and spermatozoa and pigment-forming cells. Trophoblast cells appear relatively specialized in that they are large and flattened with numerous microvilli forming the continuous close-knit wall of the blastocyst. They include about two-thirds of all the cells of the blastocyst. Electron microscope investigations show that their membranes are closely apposed and interdigitated, linked at intervals by the 'tight junctions' characteristic of closely integrated tissues. In the rat, these tight junctions have been observed as early as the 8- to 16-cell stage, around the periphery of the embryo. Functionally, the trophoblast acts first as a pump, in that it is responsible for active transfer of fluid into the blastocyst cavity. Later, the trophoblast is the first to make contact with maternal tissue, ingesting the disintegrating uterine epithelium and invading the deeper layers. In contrast, the inner cell mass consists of small, rounded, rapidly dividing cells, showing minimal mutual adhesion or specialization until after implantation has occurred. As we shall see later, recent work has uncovered some of the causal basis which underlies this first and very fundamental step in differentiation. At a later stage the endoderm differentiates, forming a layer of cells lining the inner surfaces of trophoblast and inner cell mass. In the rat, the endoderm cells are said to come from the trophoblast, while in the sheep they are believed to arise by migration or separation as a layer from the inner cell mass. In the goat and pig, larger cells at one pole can already be seen in the morula: these constitute the embryonic ectoderm, while the trophoblast develops separately and does not enclose the embryonic mass. In these forms, therefore, to speak of an 'inner cell mass' is misleading. The endoderm is believed to arise from trophoblast cells.

EXPERIMENTAL MANIPULATION

Until about 1960, very little experimental embryology had been done on mammals because of the small size and inaccessibility of mammalian embryos. During the last 10 years, these obstacles have been largely overcome: mouse embryos, and to a lesser extent rabbit and sheep embryos, have been cultivated *in vitro* throughout the cleavage period, and subjected to a variety of experimental procedures. After *in vitro* cultivation they can be transferred to the uterus of a female at an appropriate stage of pregnancy or pseudopregnancy, so that their developmental potential can be assessed.

Early claims that parthenogenetic activation of unfertilized rabbit eggs was followed by full normal development have not been confirmed. Parthenogenetic activation of the eggs of several mammalian species (sheep, rabbit, mouse) has certainly been achieved by a variety of methods, but development usually ceases at or before the blastocyst stage. Recently, however, parthenogenetic development of mouse embryos has been induced *in vivo* by electrical treatment, by Krystof

The embryo

Tarkowski in Warsaw, and *in vitro* by removal of the cumulus cells, by Christopher Graham in Oxford, and in both cases the parthenogenetic embryos developed further than the blastocyst stage. The cultured embryos showed extensive growth on transfer to sites outside the uterus, while those remaining in the uterus survived up to mid-gestation. In both series some of the embryos developed as haploids, with half the normal chromosome number, some as diploids, and some as haplo-diploid mosaics.

Triploidy has been induced in mice, rats and rabbits by suppression of second polar body formation. The triploid embryos develop apparently normally to mid-gestation, and then die. Spontaneous triploidy has been detected in mouse embryos, and is thought to constitute a significant cause of embryonic mortality in man. (Triploid development is discussed further in the last chapter of this book and of Book 4.)

The effect of destroying one or more cells of the early embryo has been investigated. When 1 cell of the 2-cell mouse embryo is destroyed, the surviving blastomere usually gives rise to a normal blastocyst, though with half the usual number of cells. After transfer to a foster-mother, development continues normally, culminating in a full-sized, fertile adult mouse. Regulation of size is achieved about half-way through gestation. When 3 cells of the 4-cell stage are destroyed, an abnormal blastocyst frequently develops, consisting of a trophoblastic shell only, with no inner cell mass. Such a structure is incapable of further development. When 7 cells of the 8-cell stage are destroyed, these abnormal trophoblastic vesicles predominate. In rabbits, on the other hand, normal young have been born after destruction of 7 out of 8 cells. Identical twinning has been achieved experimentally in mice, by separating the two blastomeres at the 2-cell stage and allowing each to develop independently in culture. Sometimes both 'twins' give rise to normal blastocysts; sometimes two trophoblastic vesicles are produced. Such twin pairs have not been grown past the blastocyst stage.

Mouse embryos can not only be taken to pieces, they can also

6

be joined together. Techniques for removal of the zona pellucida and fusion of cleavage stages (most conveniently, 8-cell embryos) were developed independently by Krystof Tarkowski, and by Beatrice Mintz in Philadelphia. The double embryo gives rise to a single blastocyst containing twice the normal number of cells, and development after transfer to a foster-mother again proceeds normally. Size regulation is complete by the 17th day of gestation; exactly when or how such regulation occurs is not yet known. The ease and consistency with which normal development takes place established that the ultimate fate of the blastomeres is in no way irreversibly determined by the 8-cell stage. The technique of embryo fusion has been used in several different laboratories to produce chimaeric mice, containing two genetically distinct cell populations, and has yielded valuable information on cell lineages in development, the genesis of coat colour patterns, phenotypic interactions between cells of different genotype, and also the differentiation of sex, a subject that is discussed further in the next chapter.

The techniques of blastomere destruction and embryo fusion can be used to throw light on the causal basis of differentiation into trophoblast and inner cell mass. Two alternative models have been proposed (Fig. 1-1). According to the first or preformationist hypothesis, two types of cytoplasm co-exist in the fertilized egg, and are separated out during cleavage. Cells that receive one type of cytoplasm differentiate as trophoblast cells, while the remainder constitute the inner cell mass. This model is hard to reconcile with the observation that the first two blastomeres, when separated, may develop into two normal blastocysts, or both may develop into trophoblastic vesicles with no inner cell mass. The alternative hypothesis is an epigenetic rather than a preformationist one, and postulates that the differences which distinguish trophoblast from inner cell mass do not stem from any preformed regional differences in the fertilized egg, but arise during development as a result of some environmental difference between cells. This difference can neither be related to the point of sperm entry, since, as we have

7

seen, mouse eggs will develop normally to the blastocyst stage after parthenogenetic activation in the complete absence of sperm participation; nor can it involve the position of the egg in the reproductive tract, since development from the 1-cell up to the blastocyst stage is possible *in vitro*. The most obvious environmental difference is that, at the morula stage, some cells are in the middle surrounded by cells on all sides, while other cells, at the periphery of the embryo, are in contact with the external milieu.

The second model is consistent with the observation that, in the mouse, destruction of all but one of the cells at the 8-cell stage usually prevents the development of any inner cell mass, since at the critical morula stage cell number in the 1/8 egg would be too small to permit any cells to be entirely enclosed. In the rabbit on the other hand, the critical stage may, like blastocyst formation, be delayed until a couple of cleavage divisions later, so that destruction of 7 out of 8 cells at the 8-cell stage may still allow some cells to be surrounded by others before the

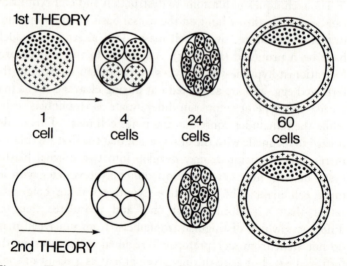

Fig. 1-1. Two theories of how cells in the outer layer of a blastocyst come to be different from those in the inner cell mass. (From A. McLaren. *Proc. Roy. Inst.* **42,** 153 (1969).)

blastocyst begins to form, and hence the development of normal blastocysts is possible. More conclusive evidence in favour of the environmental model is provided by Nina Hillman and Christopher Graham's recent demonstration that, if a radio-actively labelled 8-cell mouse embryo is experimentally sur-rounded on all sides by unlabelled embryos, by the fusion technique, then most and sometimes all of the labelled material will subsequently be found in the inner cell mass.

The inside–outside model is probably the correct explanation for differentiation into inner cell mass and trophoblast in the eggs of higher mammals with which we are most familiar. We do not know the explanation in, for instance, the yolky eggs of the marsupial native cat *Dasyurus* which consist of a single layer of cells with none in the middle, until well after the blasto-cyst is formed; nor can we apply the inside–outside explanation to the goat and pig if it is true that in these species the embry-onic cells develop at one pole of the morula and are not enclosed by the trophoblast.

The interactions between the trophoblast and inner cell mass, and their relative functions, have been greatly illuminated in the mouse blastocyst by the elegant microsurgical experiments of Richard Gardner, described in more detail in the fourth chapter in this book.

BIOCHEMICAL STUDIES

Metabolic rate, whether judged by oxygen uptake or carbon dioxide output *in vitro*, increases little during the early cleavage divisions, but rises sharply between the morula and blastocyst stage in both mouse and rabbit (Fig. 1-2). In the rabbit, this is partly due to a shift from the hexose monophosphate oxidation pathway to the more efficient Embden–Meyerhof pathway and the tricarboxylic acid (TCA) cycle.

All mammalian embryos are probably dependent on a con-tinuing supply of energy from the maternal environment, unlike frog or sea-urchin embryos, which can develop in pure water or

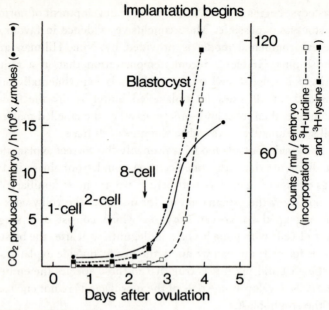

Fig. 1-2. CO_2 output and RNA and protein synthesis in the pre-implantation mouse embryo.

a simple salt solution respectively. The compounds that the mouse embryo is able to utilize as an energy source *in vitro* change markedly as development proceeds (Table 1-1). The fertilized 1-cell egg resembles the oocyte in requiring pyruvate or oxaloacetate; the 2-cell embryo can also utilize phosphoenolpyruvate and lactate; but it is not until the 8-cell stage onwards that glucose can be used as a source of energy. If follicle cells are present around the fertilized egg, development is possible from the very earliest stage with glucose as the sole energy source in the culture medium, since the maternal cells convert the glucose into pyruvate. The process of biochemical differentiation during cleavage may be related to an alteration in the ultrastructure of the mitochondria, since John Biggers and his colleagues in Baltimore have reported that 1- and 2-cell stages contain rounded mitochondria with few cristae like those of developing oocytes, while from the 8-cell stage on-

TABLE I-I. Energy sources capable of supporting the development *in vitro* of mouse eggs and embryos at different stages of development

Substrate	Oocyte	1-cell	2-cell	8-cell
Pyruvate	+	+	+	+
Oxaloacetate	+	+	+	+
Phosphoenolpyruvate	−	−	+	+
Lactate	−	−	+	+
Glucose	−	−	−	+

wards the normal adult type is present, with abundant cristae.

A large increase during cleavage in the amount of glycogen in the cytoplasm has been shown in the mouse. Whether this is used as an energy source at a later stage of development is not known.

The amount of RNA in the early mouse, rat and rabbit embryo has been measured by biochemical and cytophotometric (Azure-B staining) methods. The unfertilized egg is rich in RNA, and during the first few cleavage divisions the total amount of RNA increases little or even declines. Since *in vitro* studies on the incorporation of radioactively-labelled uridine into embryos indicate that some RNA synthesis is taking place at this time, breakdown of RNA must also occur. The rate of incorporation of uridine increases as development proceeds, but some incorporation has been demonstrated from the 2-cell stage onwards in mouse, rabbit and hamster embryos. Labelling is confined to the nucleus at the 2-cell stage, appearing for the first time in the cytoplasm at the 4-cell stage. Nuclear labelling at the 4-cell stage is concentrated over the nucleoli, suggesting that ribosomal RNA is being synthesized. For the 4-cell mouse embryo, this has been confirmed by column chromatography and the use of sucrose gradients, which show large radioactive peaks of 28S and 16S material. Also at the 4-cell stage, nucleoli of adult type are first seen. It has been calculated that by the blastocyst stage, 10^4 new ribosomes are made every minute in each cell of a mouse embryo, a rate of production comparable to that of adult cells in culture.

Similar techniques have been used to demonstrate the synthesis of low molecular weight RNA, both 5S and 4S (transfer RNA), from the 2-cell stage onwards. Synthesis of messenger RNA has not been conclusively demonstrated in early embryos.

Mammalian embryos differ strikingly from most other embryos (e.g. those of amphibians, teleost fish, echinoderms, insects) not only in requiring an external energy source, but also in the precocious onset of RNA synthesis. Both differences probably reflect the small size of the mammalian embryo and the very limited nature of its nutrient reserves. In the mouse, ribosomal and transfer RNA are produced as early as the 2-cell stage, and by the blastocyst stage the rate of synthesis of these RNA classes is indistinguishable from that of an adult cell. The frog embryo does not achieve this degree of biochemical maturity until the gastrula stage, by which time it contains 30 000 cells. Because early RNA synthesis is necessary for normal development, mammalian embryos are very susceptible to the action of actinomycin D which inhibits the formation of RNA: a 10^{-7} molar concentration in the culture medium is sufficient to reduce RNA synthesis in the mouse by 90 per cent. By contrast, amphibian and echinoderm embryos, in which RNA synthesis is so much more retarded, are highly resistant to the harmful effects of actinomycin D during cleavage and blastulation; throughout these stages they depend upon RNA elaborated in the course of oogenesis.

Protein synthesis, as judged by the incorporation of radioactively labelled aminoacids, can be detected in cultured mammalian embryos from the fertilized egg onwards. *In vivo*, however, very little protein synthesis can be demonstrated during early cleavage. As with RNA, since the total amount of protein has been shown in the mouse embryo to decline during the first few cleavage divisions, breakdown of protein must occur to a greater extent than synthesis. The rate of protein synthesis increases dramatically from the morula to the blastocyst stage, possibly coinciding with the first major use of RNA templates from the embryo's own genome.

Little is known about the synthesis of individual proteins during the first few days of life. Spindle proteins, necessary for the formation of the mitotic apparatus, are probably produced at this time since cleavage can be prevented by treatment with a specific protein synthesis inhibitor, puromycin. The activities of some enzymes during the pre-implantation period have been determined by Ralph Brinster in Philadelphia and by Charles Epstein and his colleagues in San Francisco. Lactate dehydrogenase (LDH) is present in mouse embryos in large amounts during the first two days of development, and thereafter declines exponentially, presumably owing to selective degradation. The sex-linked enzyme glucose-6-phosphate dehydrogenase (G6PD) (the gene for which is carried on the X-chromosome) shows the same pattern of activity. Other enzymes, including malate dehydrogenase (MDH), isocitrate dehydrogenase (IDH) and an aldolase, show a modest increase in activity, about 30 per cent in 4 days, which is probably due to a low rate of enzyme synthesis. Hexokinase and two phosphoribosyltransferases show a marked increase in activity during this period. The pattern of activity appears to be similar *in vivo* and *in vitro*.

There is some evidence that characteristic proteins identified as transplantation antigens may be expressed on the surface of the blastocyst, but that these reflect the embryo's own genotype rather than that of the mother has not yet been established.

GENE ACTION

An important problem, not yet resolved, concerns the stage at which the embryonic genome first becomes active. At what stage is the chromosomal DNA of the embryo first transcribed?

The embryo

At what stage is embryo-derived messenger RNA first used in protein synthesis, as opposed to long-lived messenger RNA produced during oogenesis? At what stage are paternal genes first expressed? These three questions may turn out to have different answers.

The evidence, summarized above, on RNA synthesis during early cleavage of mouse embryos suggests that at least those regions of embryonic DNA that code for the ribosomal and transfer RNA must be functional from at least the 4-cell stage.

Treatment of early cleavage mouse embryos *in vitro* with a concentration of actinomycin D (10^{-7} M) sufficient to reduce RNA synthesis by 90 per cent and to depress cleavage, only reduces protein synthesis by 50 per cent. This could be interpreted to mean that by about the 4-cell stage, 50 per cent of the protein synthesis that is taking place still depends upon long-lived messenger RNA synthesized during oogenesis or immediately after fertilization, while the other half depends upon newly synthesized messenger RNA. However, actinomycin D is known to exert a variety of effects on the functioning of the cell, and it is possible that its effect on protein synthesis in the mouse embryo operates in some way other than by inhibiting the transcription of messenger RNA. No other biochemical evidence exists as to how early in development newly transcribed messenger RNA is used in protein synthesis, but large numbers of new ribosomes begin to accumulate after the 8-cell stage. Mouse and rat chromosomes are thin and not much spiralized during cleavage, suggesting inactivity, and it is not until the blastocyst stage that they cease to differ morphologically from those of the adult.

The genetical evidence as to when paternal genes are first expressed is conflicting. On the one hand there is a much-quoted observation by Wesley Whitten and Charles Dagg in Bar Harbor, that cleavage rate of eggs of one mouse strain depends on the genotype of the father from the 2-cell stage onwards, and a suggestion by Trevor Morris at Harwell that eggs lacking an X-chromosome only survive the first cleavage division if fertilized by an X-bearing rather than a Y-bearing spermatozoon. On the other hand there is a considerable body of evidence to suggest that chromosomal deficiencies or incompatibilities only make themselves apparent at the blastocyst stage, when cleavage has achieved a more normal nucleo-cytoplasmic ratio. No genes are known which affect mammalian embryos during early cleavage. The earliest acting lethal gene in the mouse, t^{12}, does not affect cleavage, but causes disturbed RNA synthesis and nucleolar development in the late morula. Inter-species hybrids in which fertilization occurs usually undergo normal cleavage, only to fail at the blastocyst stage, e.g. rabbit (*Oryctolagus cuniculus*) × hare (*Lepus europaeus*) and rabbit × cottontail rabbit (*Sylvilagus floridanus*). In rats, Andrzej Dyban and his colleagues in Leningrad have induced chromosome aberrations by X-irradiation of the male gametes or treatment of the zygotes with antimetabolites, and obtained embryos with varying degrees of aneuploidy, i.e. loss of one or more chromosomes. Even where the missing chromosome was one of the largest of the set, cleavage and blastocyst formation occurred normally, the embryos usually dying at about the time of implantation. The same result was obtained by Bunny Austin at Covington with aneuploid rabbit embryos.

The balance of evidence suggests that, though RNA and protein are synthesized by the embryo from a very early stage, the embryonic genome does not begin to control the course of development until blastulation. Verne Chapman and his colleagues have recently found that mouse embryos from crosses between strains characterized by different electrophoretic variants of the cellular enzyme glucose phosphate isomerase first show the paternal variant at the late blastocyst stage.

Sex chromatin, the visible consequence of the inactivation of one X-chromosome in each cell of the female mammal, is not seen during cleavage. This is consistent with the fact that oocytes show double the normal activity of the X-linked enzyme G6PD during oogenesis; at this time both X-chromosomes appear to be functioning, and from the oogonial stage onwards neither is inactivated (see Book 1, Chapter 2). Sex chromatin has been reported in the Syrian hamster as early as the 8-cell stage, in the rabbit on the 4th day of pregnancy at about the 200-cell stage, in the monkey and human at about the time of implantation, in the cat at the late blastocyst stage, and in the rat not until 2 days after implantation (7th day of gestation). Whether the stage of development at which the surplus X-chromosome is inactivated bears any relation to the stage at which the genome is activated, is not known.

The biochemistry of the fluid in the rabbit blastocyst cavity has been studied with radioactive tracers to determine the passage of substances into and out of the blastocyst. Before implantation the fluid is very rich in potassium and bicarbonate, which appear to be actively drawn into the blastocyst from the uterine fluid. As implantation proceeds, potassium and bicarbonate fall to the levels found in the maternal serum, while protein and glucose, previously present in small amounts only, increase up to maternal serum levels. Phosphorus and chlorides also increase greatly in concentration. Thus the blastocyst possesses a high degree of metabolic selectivity, actively controlling the rate of entry of substances from the surrounding uterine fluid.

IN VITRO CULTURE

Much of the information summarized above on the biochemistry of the pre-implantation embryo could only have been obtained by culturing the embryos *in vitro*, outside the maternal body, in a chemically defined medium. Since such a medium contains no complex constituents, for instance serum or plasma, the nutritional requirements of the embryo can be analysed fairly easily. The first widely used chemically defined medium, devised by Wesley Whitten, consisted essentially of a bicarbonate-buffered salt solution (Krebs–Ringer) supplemented with glucose and albumin. It supported the development of mouse embryos from the 8-cell to blastocyst stage. By subsequent transfer to foster-mothers, the author in collaboration with John Biggers showed that the developmental potential of the cultured embryos was unimpaired, in that they gave rise to normal fertile adult mice.

The period over which embryos could be successfully cultured was pushed backwards in time by Ralph Brinster, who modified the medium, principally by the addition of lactate and pyruvate, so as to allow the development of 2-cell eggs which cannot utilize glucose as an energy source (see Table 1-1). Brinster also investigated the optimal range of pH and osmolarity. For several years progress was held up because development *in vitro* appeared to be blocked at the 2-cell stage. Fertilized 1-cell mouse eggs would undergo one cleavage division and then stop unless they were returned to the reproductive tract, although late 2-cell eggs would develop *in vitro* up to the blastocyst stage. The 2-cell block did not show itself if embryos were placed inside the ampullary region of the oviduct growing in organ culture in a chemically defined medium. This suggested that the ampulla provides a specialized environment in some way necessary for the early development of the mouse. Recently, however, embryos of particular genetic constitution, in certain laboratories, have been cultured all the way from 1-cell to blastocyst *in vitro*, but the difficulties encountered in repeating these experiments in other laboratories suggest that we are still ignorant of some of the essential requirements of the culture system.

The albumin in the culture medium is probably fulfilling a physical rather than a nutritional role, e.g. acting as a membrane stabilizer, since it can be replaced by a synthetic macromolecule, polyvinylpyrrolidone (PVP), permitting development of mouse embryos from the 2-cell up to the blastocyst stage in the absence of any external source of nitrogen.

Rat and rabbit embryos have also been cultured in chemically defined medium during the pre-implantation period. One of the protein fractions present in the rabbit uterus shortly before implantation (blastokinin, uteroglobin) may be required for the expansion of the rabbit blastocyst, but recently expansion has been successfully achieved *in vitro* in a very simple medium.

Other mammalian embryos (e.g. pig, sheep, human) have been successfully cultured during the pre-implantation period in various media, usually containing serum, but in no species have the nutritional requirements been so precisely analysed as in the mouse.

Once the blastocyst stage has been achieved, mouse embryos have been shown by time-lapse cinematography to undergo striking cycles of contraction and expansion. A rapid contraction, thought to be mediated by microfibrils in the outer trophoblast layer, reduces the blastocoele cavity to about one third of its former size, and is followed by a slower expansion phase brought about presumably by an active pumping of the vessel into the blastocoele cavity. Whether this cycle also takes place *in vivo*, and if so what is its function if any, remains unknown. If serum is added to the culture medium, the mouse blastocyst attaches to the glass or plastic surface of the culture medium, and the cells grow out over the surface as a two-dimensional sheet. Rabbit blastocysts will continue developing up to the stage of embryos with beating hearts if serum is added to the medium. The presence of a 'feeder' layer of cells on the bottom of the culture dish also aids development.

The process of implantation has not yet been simulated *in vitro*, though Anthony Glenister in London has described rabbit blastocysts attaching to a piece of cultured uterine endometrium and continuing their development up to the beating-heart stage. Mouse, rat and hamster embryos have been dissected out of the uterus several days after implantation, and have been grown in circulating medium or on plasma clots, usually by the technique of Denis New, up to the late somite stage. The presence of serum in the medium seems essential.

TWINS

Twins can be monozygotic (one-egg, identical) or dizygotic

(two-egg, fraternal). Dizygotic twins are formed when two eggs are shed in a single ovulation period, and fertilized by two separate spermatozoa. The resulting young resemble each other genetically no more than do any other brothers and sisters. Indeed, the difference in weight between male and female lambs is greater in male–female twin pairs because of embryonic competition, than when male and female lambs are born in like-sexed twin pairs. In species that normally produce several young at a time (polytocous, litter-bearing), the concept of dizygotic twinning scarcely applies.

Injection of gonadotrophic hormones can induce multiple ovulations and hence promote dizygotic twinning. This technique finds practical application in cattle breeding, while in women multiple births are an undesirable side-effect of the use of gonadotrophins to combat sterility. The incidence of spontaneous dizygotic twinning varies widely among different populations as it is affected not only by genetic but also by dietary and other environmental factors.

When two or more embryos are implanted in the same uterus, they may share a common blood circulation (placental anastomosis), with the consequence that hormones and circulating cells are transferred between embryos. In cattle, when twinning occurs it is almost always accompanied by placental anastomosis: twin calves therefore share each other's blood groups and can accept skin grafts from one another, and if they are of opposite sex, the sexual development of the female (freemartin) is impaired (see Chapter 2 in this book). In contrast, marmosets almost invariably produce dizygotic twins and, although exchange of blood cells occurs, no abnormalities of sexual development have been observed.

Monozygotic twins originate from a single fertilized egg, and hence resemble one another very closely in all genetically determined characters. For instance, they are always of the same sex. The only known exceptions are a few cases reported in our own species where loss of a sex chromosome from one twin early in development has led to the birth of XO (sterile female) and

XY (normal male) 'identical' twins (see Chapter 2). The incidence of monozygotic twinning is much lower than that of dizygotic, and varies very little among different populations. In cattle, up to 5 per cent (depending on breed) of all births are twins, but only about 0.1 per cent of births are monozygotic twins.

Many instances of monozygotic twinning probably originate after implantation. A single blastocyst implants, and the single inner cell mass then differentiates into two primitive streaks, giving rise to two separate individuals. The classic example of this phenomenon is seen in the nine-banded armadillo, where the single blastocyst invariably produces four primitive streaks, so that the pregnancy invariably results in four monozygotic young. Alternatively, the inner cell mass of a single blastocyst may duplicate at an earlier stage than implantation, a condition that has been reported in both sheep and pigs.

EMBRYOS IN THE REPRODUCTIVE TRACT

After fertilization in the ampullary region of the oviduct (Fallopian tube), the mammalian egg is transported down the oviduct and into the uterus (Fig. 1-3). In some marsupials (opossum, rat kangaroo) this may take 24 h or less, while the embryos of some carnivores may stay in the oviduct for up to a week, but in most mammals the journey from ampulla to uterus takes 2–4 days (Table 1-2). The rate of transport down the oviduct is not uniform: in some species the embryos remain for a long time at the ampullary-isthmic junction, halfway down the oviduct, while in others they are held up at the entrance to the uterus, the utero-tubal junction.

It is essential for the survival of the embryo that the oviduct should not be traversed too rapidly. At oestrus, the distended uterus provides a favourable environment for spermatozoa, but a very unfavourable one for eggs; not until at least a day later does the uterine environment become tolerable for the embryo. One way in which oestrogen exerts a contraceptive effect is by hastening the passage of embryos through the oviduct and into the toxic surroundings of the early post-oestrous uterus.

The mechanism of embryo transport is thought to involve a combination of ciliary action and muscular peristalsis. Only the epithelium of the upper part of the oviduct is ciliated, so ciliary action may be mostly concerned with passage of the unfertilized egg into the ampulla after ovulation, and with maintaining fluid currents in the ampulla during the fertilization period. In the Rhesus monkey, the cilia of the oviduct epithelium atrophy during the later phase of each cycle, and are reconstituted early in the next under the influence of oestrogen. The zona has recently been shown to be essential for the transport of cleavage stages in the mouse, although blastocysts experimentally injected into the top of the oviduct are transported to

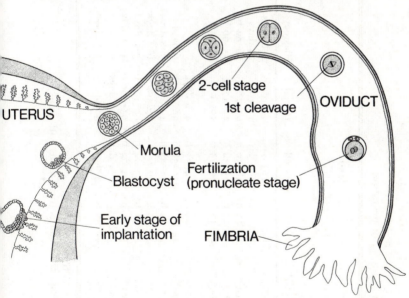

Fig. 1-3. Development of the human embryo in the reproductive tract, from fertilization to implantation. (After H. Tuchmann-Duplessis, G. David and P. Haegel. *Illustrated Human Embryology*, vol. 1. Springer-Verlag, New York; Chapman and Hall, London; Masson et Cie, Paris (1971).)

TABLE 1-2. Timing of gestation

Animal	2-cell	4-cell	16-cell	Blastocyst	Entry into uterus (days)	Implantation (days)	Gestation (fertilization to birth) (days)
Man*	1–2 days	2–3 days	3–4 days	4–6 days	3	8–13	252–74
Rat†	1–2 days	2–3 days	4 days	4½ days	3	5	20–22
Mouse†	21–23 h	38–50 h	60–70 h	66–82 h	3	4	19–20
Rabbit†	21–25 h	25–32 h	40–47 h	75–96 h	2.5–4	7–8	30–32
Guinea pig†	23–48 h	30–75 h	107 h	115 h	3.5	6	63–70
Rhesus monkey*	26–49 h	24–52 h	4–6 days	—	3	9–11	159–74
Horse*	24 h	30–36 h	98–100 h	6 days	4	28	335–45
Cow*	27–42 h	50–83 h	4 days	7–8 days	3–4	30–35	275–90
Sheep†	38–39 h	42 h	3 days	6–7 days	2–4	15–16	145–55
Pig†	25–51 h	25–74 h	80–120 h	5–6 days	2–2½	11	112–15
Ferret†	51–71 h	64–74 h	95–120 h	4½–6 days	5–6	7–8	42
Mink†	3 days	3–4 days	5–6 days	—	8	25	42–52
Cat†	40–50 h	3 days	4 days	5–6 days	4–8	13–14	52–65

* Times from ovulation.
† Times from coitus (first coitus with mink).

ne uterus whether or not the zona pellucida is present. Possibly the muscular peristalsis which squeezes the embryos down the oviduct would disrupt and destroy any group of cells not surrounded either by a zona, or by a closely integrated tissue like the trophoblast. Thus, one important function of the zona pellucida is to hold the cleaving egg together during its passage down the oviduct and to prevent it sticking to the oviduct walls.

Once the embryos have entered the uterus, a period of at least a day (in mouse or rat) and several weeks in the horse, intervenes before implantation begins. Muscular action is again important in species with more than one young per litter (polytocous), in order to spread the embryos along the length of the uterus, and thus minimize crowding which in later pregnancy might lead to embryonic death. Philip Dziuk and his colleagues at Urbana have examined in detail the timing of distribution of embryos entering the pig uterus from one side only, and find that the proportion of the total uterus reached by such embryos increases from 13 per cent on the 6th day to 86 per cent on the 12th day. Where the two uterine horns open separately into the vagina, as in the rat, the eggs from each ovary are confined to the corresponding uterine horn. In other species, e.g. the pig, the two uterine horns are to some extent connected, so that migration of embryos from one horn to another is possible. This is termed 'internal migration', in contrast to 'external migration', which is the migration of an unfertilized egg from the ovary on one side into the top of the oviduct on the other side, in species lacking a closed ovarian capsule.

In the mouse, internal migration can occur but in undisturbed pregnancies is rare, affecting less than 1 per cent of embryos. In the pig, on the other hand, it occurs very frequently. From data on the relative distributions of corpora lutea and implantation sites between the two sides of the reproductive tract, it has been calculated that an egg shed from one ovary has an equal probability of implanting in the uterine horn on the opposite side as on the same side of the tract. In consequence, although the numbers of eggs shed from the two ovaries are negatively correlated, the numbers of embryos implanted in the two horns show a strong positive correlation, and indeed seldom differ by more than one. In intersex pigs, which have a functional ovary on one side and a testis or ovotestis on the other, all the eggs are shed on the ovarian side, but most of the implantation sites are found in the uterine horn on the opposite side. This suggests that the testis tissue exerts a local hormonal effect on the uterine horn of that side, interfering with the normal pattern of migration of embryos between sides.

In the mouse, the blastocysts are not only spread down the length of the uterine horns by the muscular action of the uterus, but are positioned by the same process along the anti-mesometrial surface of the endometrium, i.e. on the wall opposite to that by which the uterus is suspended in the peritoneal cavity. The role of the blastocyst must be a passive one, since beads or small pieces of muscle placed in the uterus are treated in the same way.

IMPLANTATION

Throughout the animal kingdom, embryos of related groups tend to resemble one another closely, more indeed than do the corresponding adult forms. The exceptions occur when embryonic adaptations to particular environmental conditions evolve, as with the diverse spines and processes of sea urchin larvae. In this light we must view the great diversity of modes of implantation and placentation seen in different mammalian groups, in contrast to the striking similarity of cleavage stage embryos in mammals. We are largely ignorant of the forces of the maternal environment which have evoked these various forms of implantation.

The embryo is said to be implanted or attached when it becomes fixed in position, and establishes physical contact with the maternal organism. The embryonic tissue specialized for interaction with the uterus is the trophoblast. Characteristic trophoblastic giant cells, many times the size of normal cells and often highly invasive, develop during the course of implantation, and at a later stage many of the cell boundaries break down to give a trophoblastic syncytium, with many nuclei contained in a single mass of cytoplasm. The manner in which the trophoblast interacts with the maternal tissues largely determines the morphological form of the mature placenta. For instance, in the cow, sheep, horse and pig, the blastocyst remains in the uterine lumen, and the trophoblast (or chorion, as it is called later in pregnancy) makes close contact with the uterine epithelium. Such placentae are syndesmochorial or epitheliochorial. In the endotheliochorial placenta characteristic of carnivores, the uterine epithelium is broken down, and the trophoblast, in the form of chorionic projections or villi, invades the deeper uterine tissue, the stroma, until it reaches the endothelium lining the maternal blood vessels. The most intimate contact between the embryo and the maternal organism is achieved in the haemochorial placenta, found in rodents and in our own species, where the chorionic villi proceed to break down the blood vessels until they are bathed in maternal blood. In general, those species in which the embryo becomes more intimately connected with the mother are also those in which implantation starts earlier.

Horse, cow, sheep, pig

The horse blastocyst, after the loss of the zona pellucida, becomes distended by the fluid pressure in the blastocoele to a diameter of more than 5 cm, and remains pressed against the uterine epithelium for nearly 2 months without forming its final attachment to the uterus. During this period it is nourished by uterine milk (histotrophe), a mixture of uterine secretion and tissue debris. At the end of the 3rd week, specialized groups of cells appear on the trophoblast which facilitate ingestion of the uterine milk, and during the next few weeks form a transient attachment to the uterine epithelium. Not until the 10th week do chorionic villi grow out into the folds of the uterine wall, and by the 14th week attachment is complete.

Cows and sheep are distinguished by the permanent presence of specialized attachment areas in the uterus, the caruncles. The blastocysts are more elongated than in the horse, and attachment occurs earlier, beginning in the 3rd week of gestation in sheep, and in the 5th week in cows. Trophoblast invades the caruncles, to form areas of intimately interdigitated embryonic and maternal tissue, the cotyledons, which nourish the developing embryo (Fig. 1-4).

In the pig, the blastocysts elongate to a much more extreme degree, so that by the time migration around the uterus is complete, they may measure more than a metre in length, with a small central embryonic area or inner cell mass, and extended ribbons of trophoblast applied to the folds of the uterine endometrium. In the immediate pre-implantation period, pig embryos suffer extremely heavy mortality. By about the 13th day of gestation, small areas of attachment have appeared between the trophoblast and the uterine epithelium; the uterine epithelium is modified where it is in contact with the trophoblast, and a dense network of capillaries forms below it. During the following week attachment is completed and the trophoblast becomes modified for absorption.

In all these species, implantation is a very gradual process, allowing much scope for controversy as to when it begins and ends, while the attachment which is achieved between trophoblast and endometrium is relatively tenuous, at least in the early stages. Since a period of weeks may elapse before any appreciable contact occurs between embryonic and maternal tissues, the embryo depends for its nourishment on absorption from the surrounding uterine fluid.

Man

By the end of the first week of pregnancy, the human blastocyst has already entered the uterus, lost its zona pellucida, stuck to

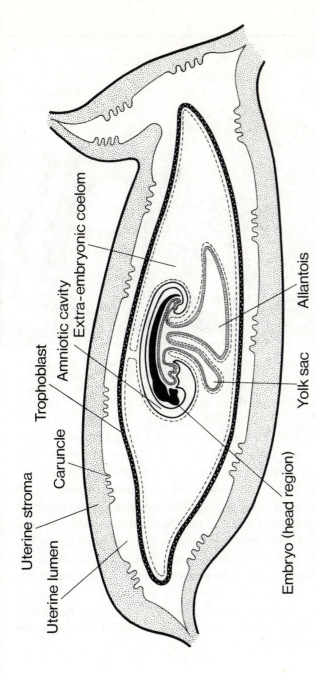

Uterine stroma

Uterine lumen

Caruncle

Trophoblast

Amniotic cavity

Extra-embryonic coelom

Embryo (head region)

Yolk sac

Allantois

Fig. 1·4. Diagram to show the relation between the fetal membranes and uterine cavity at the beginning of implantation in the sheep. (From J. D. Boyd and W. J. Hamilton. In *Marshall's Physiology of Reproduction*, vol. 2, Ed. A. S. Parkes. Longmans Green (1952).)

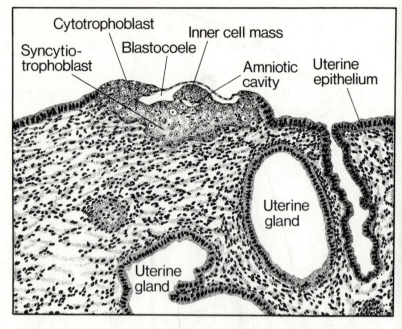

Fig. 1-5. Section of a 7½-day human embryo implanting in uterine endometrium. The thin membranous portion of the blastocyst wall has collapsed. (From W. J. Hamilton, J. D. Boyd and H. W. Mossman. *Human Embryology*. Heffer, Cambridge (1945).)

the uterine wall, broken down the epithelium and sunk through into the underlying stroma (Fig. 1-5). There is no uterine milk, because there is no need for it. The trophoblast proceeds to invade the wall, forms a syncytial layer in which adjacent cell boundaries are lacking, and engulfs the maternal blood vessels. A layer of specialized decidual cells forms in the uterine stroma. As the trophoblast swallows up more blood vessels (Fig. 1-7), a system of sinusoids develops, spaces filled with maternal blood and forming the basis of the maternal circulation in the placenta. Trophoblastic processes project into the sinusoids and become vascularized by mesodermal outgrowths from the developing embryo, forming the basis of the embryonic circulation in the placenta.

Mouse

Mouse and rat resemble man, and differ strikingly from the other species we have considered, in the early stage of gestation at which implantation begins, the rapidity with which it proceeds, and the intimate contact between embryo and uterus that is achieved. They differ from man in having a more precocious and extensive decidual reaction in the uterus.

At the end of the 4th day of pregnancy, the mouse blastocysts lie along one extremity of the slit-like uterine lumen, with the walls of the lumen pressed tightly around them. The zona pellucida is dissolved by an enzyme produced by the uterus, and simultaneously the first signs of a reaction occur in the uterine stroma in the neighbourhood of the blastocyst (Fig. 1-6). Capillary permeability undergoes a sharp local increase, and the stroma becomes swollen with fluid. Electron microscope studies on a number of species have shown that, up to this stage of pregnancy, the surfaces of both trophoblast and uterine epithelium are covered with numerous interdigitated micro-villi, and separated by a gap of about one micron; once attachment is under way the microvilli disappear and the embryonic and uterine surfaces become very closely apposed. The uterine epithelium surrounding the blastocyst begins to degenerate and at the same time the trophoblast invades between the epithelial cells, engulfing the dead cells, and pushes into the stroma. The stromal cells undergo the decidual reaction, increasing both in size and in number until they constitute a thick, closely-knit layer enclosing the developing embryo on all sides (Fig. 1-6). During the first few days following implantation, the embryo is nourished by the absorptive activities of the trophoblast; later a true placenta is formed involving both decidua and trophoblast.

Much of what we know about the hormonal control of implantation, as well as the causal interactions between embryo and uterus during the implantation process, has been derived from experiments on laboratory animals, especially rats and

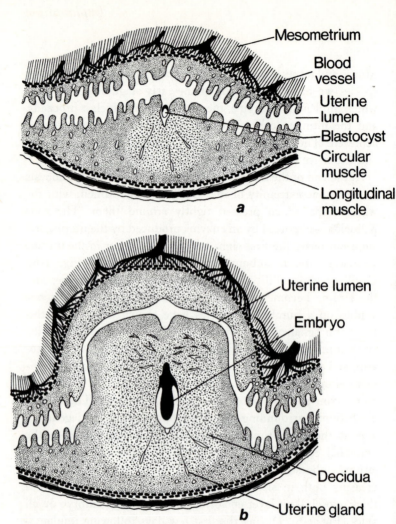

Fig. 1-6. *a* Longitudinal section through the mouse uterus about 5 days after mating. A blastocyst is implanted at the anti-mesometrial side of the uterine lumen and has induced an early decidual reaction. *b* Longitudinal section through implantation site about 7 days after mating in the mouse. The decidual response has developed further and the uterine lumen is almost occluded.

(From G. D. Snell and L. C. Stevens. In *Biology of the Laboratory Mouse*. Ed. E. L. Green. 2nd edition. McGraw-Hill, New York (1966). Copyright 1966 McGraw-Hill Book Company. Used with permission.)

mice. Since implantation is so varied in different groups, extrapolation from one species to another is dangerous. For example, implantation in the hamster seems not to require oestrogen, and the same is probably true of the guinea pig. We shall describe briefly the situation as it is at present understood in the mouse. (Further information on the role of hormones in pregnancy will be found in the fourth chapter of Book 3.)

The uterus is prepared for implantation by the action of the ovarian hormones progesterone and oestrogen. When the uterine endometrium is in an appropriately sensitized condition, a decidual reaction can be induced not only by the blastocyst as in a normal pregnancy, but by an artificial stimulus, such as the injection of oil or air, or scratching with a needle. The experimentally induced decidual reaction, known as a deciduoma, resembles the spontaneous one very closely, suggesting that the blastocyst supplies only the initial stimulus for decidualization and takes no part in the subsequent development of the reaction. Degeneration of the uterine epithelium, for instance, cannot be due to the direct action of the invading trophoblast cells, since it occurs equally in deciduoma formation. Experiments in which the ovaries have been removed, progesterone and oestrogen injected in varying combinations, and finally the degree of sensitization of the uterus ascertained by testing its capacity for deciduomal formation, have done much to elucidate the hormonal sequence which acts on the uterus during normal pregnancy. For optimal uterine sensitization, a high level of oestrogen at the time of oestrus seems to be necessary, followed by a period of several days when progesterone, produced by the developing corpora lutea, is the dominant hormone; the immediate stimulus for sensitization is given by a very small amount of oestrogen which acts on the uterus on the morning of the 4th day of pregnancy. Implantation can be prevented by removing the ovaries a few hours before this time, or by removing the pituitary or treating the mouse with antibody against luteinizing hormone at a slightly earlier stage. These observations suggest

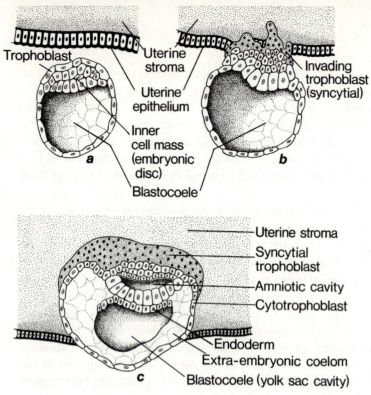

Trophoblast

Uterine stroma

Uterine epithelium

Inner cell mass (embryonic disc)

Blastocoele

a

Invading trophoblast (syncytial)

b

Uterine stroma

Syncytial trophoblast

Amniotic cavity

Cytotrophoblast

Endoderm

Extra-embryonic coelom

Blastocoele (yolk sac cavity)

c

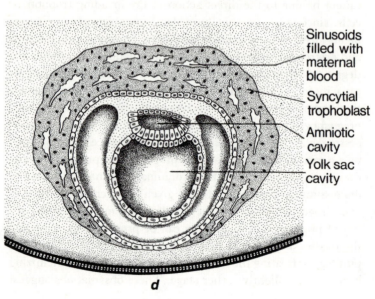

Sinusoids filled with maternal blood

Syncytial trophoblast

Amniotic cavity

Yolk sac cavity

d

32

that the oestrogen needed for implantation is produced in ovarian follicles which in turn are stimulated by pituitary hormones.

Biochemical studies on the uteri of ovariectomized rats treated with oestrogen have shown that oestrogen molecules rapidly attach to cytoplasmic receptor sites in the uterus, and pass from the cytoplasm into the nucleus, where they are associated with the chromatin. Within a few minutes of oestrogen injection, RNA synthesis in the uterus is stimulated, and within an hour the synthesis of a specific uterine protein is induced. The RNA synthesis is presumed to be the primary response, since it is not blocked by treatment with protein synthesis inhibitors such as puromycin.

In a uterus that is oestrogen-sensitized, the blastocysts become activated, the rate of RNA and protein synthesis increases, and enlargement and outgrowth of the trophoblast cells take place. If the uterus has not been sensitized, either because the pregnant mouse is suckling a previous litter, or because the ovaries have been surgically removed earlier in pregnancy, the blastocysts enter a state of dormancy or delay, and implantation is postponed. Delay of implantation during lactation occurs in mice and rats, and is also widespread among marsupials where the period of gestation is very short. It ensures that the next litter will not be born until the earlier one has been weaned. A more extended delay of implantation (embryonic diapause) occurs in other groups, e.g. badgers, bears, seals, roe deer and long-

---

Fig. 1-7. Implantation of the human embryo.
a The blastocyst is not yet attached to the uterine epithelium.
b The trophoblast has penetrated the epithelium and is beginning to invade the uterine stroma.
c The blastocyst has sunk further into the stroma and the amniotic cavity has appeared.
d The uterine epithelium has grown over the implantation site, so that the blastocyst is entirely enclosed in maternal tissue, and irregular spaces, the sinusoids, filled with maternal blood, have appeared in the syncytial trophoblast. (After H. Tuchmann-Duplessis, G. David and P. Haegel. *Illustrated Human Embryology*, vol. 1. Springer-Verlag, New York; Chapman and Hall, London; Masson et Cie, Paris (1971).)

tailed weasels, as a device to extend the gestation period so that the young will be born at a time of year optimal for survival.

We do not know what stimulus brings about activation of the mouse blastocyst in an oestrogen-sensitized uterus, as opposed to dormancy when oestrogen is lacking. Evidence suggests that the oestrogen does not affect the blastocyst directly, but induces the uterus to produce some substance that then acts on the blastocyst, either to increase metabolic rate directly or to remove some inhibitor. One substance known to be produced only by the oestrogen-sensitized uterus is the enzyme responsible for dissolving the zona pellucida: in delay of implantation the zona may be shed, but it never undergoes lysis.

The activated blastocyst in turn has an effect upon the uterus, stimulating the uterine endometrium in its immediate vicinity to undergo the decidual cell reaction which is essential if implantation is to occur. It is not the physical presence of the blastocyst which stimulates the uterus, since glass or plastic beads of a similar size have no effect; presumably therefore it is some by-product of the heightened metabolism of the embryo which acts upon the adjacent uterine epithelium. The decidualizing effect of traumatic stimuli may stem from the tissue damage which they produce.

Thus the process of implantation, at least in the mouse and probably in other species, involves a complex interaction between embryo and uterus, with each partner providing stimuli essential for the further development of the other. It is therefore not surprising that synchronous timing of embryonic and uterine development is vital if implantation is to be successful.

EMBRYO TRANSFER

The importance of synchrony was first clearly demonstrated by experiments involving surgical transfer of embryos from a donor female at one stage of pregnancy to the uterus of a recipient female at a different stage. Embryo transfers give a high success rate if donor and recipient female are at the same stage of development, or better still (since transfer may slightly retard embryonic development) if the donor is somewhat more advanced in development than the recipient. If the embryo is 'younger' than the recipient uterus, so that it is not ready to implant at the moment when the uterine endometrium is ready, the success rate is very low. Synchronization has been shown to be necessary in rabbits, mice, rats, sheep and cows. Some of the embryonic mortality that occurs in natural pregnancies around the time of implantation is probably due to faulty synchronization.

The first successful embryo transfer was carried out in 1890 by Walter Heape in Cambridge, using rabbits, but not until well into the twentieth century did the potentialities of the technique, both as a research tool and as a practical aid to animal breeding, begin to be realized.

Experimental studies which require the transfer of embryos from one female to another are of three types. Studies involving the manipulation of pre-implantation embryos *in vitro* depend upon the transfer technique for establishing the developmental potential of the experimental embryos. We have already described experiments in which several blastomeres of a cleaving embryo have been destroyed, or two embryos fused to form one, and the product reared to adulthood via the uterus of a foster-mother. Such studies are primarily embryological. Physiological studies have exploited the transfer technique to investigate problems like the effect of crowding of embryos on embryonic and placental growth and on embryonic mortality, the effect of transferring relatively advanced embryos on the development of their younger foster-siblings, and the effect of young of varying conceptual age on the timing of parturition. Thirdly, in the field of genetics, embryo transfer is used to analyse the phenomenon of maternal inheritance. When genetically distinct breeds or strains of animals are crossed and the progeny inherit certain maternal rather than paternal characteristics, the maternal influence may be exerted either through the cytoplasm of the

35

The embryo

egg or through the uterine environment. Transfer of embryos
between the two breeds distinguishes between these possibilities.
Thus, when large and small breeds are crossed, the young tend
to resemble the mother in size; in sheep, rabbits and mice,
embryos have been transferred between large and small females,
and in every case the foster mother has exerted a strong effect on
the birth weight of the transferred young, showing that it is the
uterine environment and not the egg cytoplasm that influences
growth. More surprisingly, the number of lumbar vertebrae in
mice, a character that varies between inbred strains and that
shows strong maternal inheritance, has in the same way been
shown to depend on the uterine environment rather than on the
cytoplasm of the egg. Indeed, no instances of cytoplasmic inheritance have so far been established in mammals.

Potential practical uses of embryo transfer resemble those of
artificial insemination. Reproduction at a distance is facilitated:
fertilized rabbit eggs have been flown across the Atlantic in a
vacuum flask, sheep embryos have been flown from England to
South Africa stored in the oviduct of a live rabbit, and in each
case the embryos have developed successfully to term after
transfer to a foster-mother. Increased contribution to future
generations of a cow or ewe of high genetic worth would be
another consequence of economic value if non-surgical embryo
transfer could be used in conjunction with induced ovulation of
exceptionally large numbers of eggs (see Book 5, Chapter 1):
in cattle some problems remain to be overcome before embryo
transfer becomes routine, but in sheep the first trials have already
taken place. In women, the transfer to the uterus of an egg taken
from the patient's own ovary or from that of a donor, and
fertilized *in vitro*, might alleviate some conditions of infertility,
and is likely to become technically possible within the next
decade or so (see Book 5, Chapter 4).

ECTOPIC PREGNANCIES

In a normal pregnancy, the embryos implant in the uterus. In

36

man, the posterior wall of the uterus is the usual location. When implantation occurs elsewhere, development is often abnormal. Hertig reported three abnormal and twelve normal embryos implanted on the posterior wall, six abnormal and only five normal on the anterior wall of the uterus. Implantation near the cervix leads to the condition known as placenta praevia, with possible complications at parturition.

Sometimes implantation takes place outside the uterus, leading to an ectopic pregnancy. A recent survey in the U.S. reported that 1–2 per cent of all pregnancies were ectopic; of these 96 per cent occur in the oviduct (Fallopian tube), more commonly on the right side than on the left. Since rupture of the Fallopian tube and massive haemorrhage may result from a tubal implantation, in some populations this constitutes a major reproductive hazard. Occasionally a human embryo may implant on the ovary, gut, or on an abdominal mesentery, but the chances of normal development are no better in these sites than in the oviduct. However, embryos implanted outside the uterus sometimes develop to term, in which event they require of course to be delivered surgically.

The high incidence of tubal pregnancies in man is particularly striking since, except for one or two cases in monkeys, spontaneous ectopic pregnancies are rarely if ever found in animals. Experimentally, on the other hand, mouse embryos have been transferred to a variety of ectopic sites (Fig. 1-8). In the peritoneal cavity the incidence of implantation is low, but in the kidney or in the scrotal testis, 80–90 per cent of transferred blastocysts develop. The most advanced embryo yet reported in such a site resembled a uterine embryo of about 12 days' gestation. Some situations (for instance the cryptorchid testis) seem so encouraging to trophoblast growth that further development of the embryo itself is inhibited.

The fact that mouse blastocysts survive and develop successfully in the testis emphasizes one of the most striking features of ectopic pregnancy, namely that it appears to be entirely independent of the hormonal status of the host animal. Pregnant

The embryo

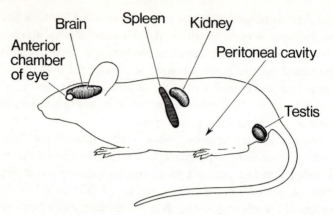

Fig. 1-8. Extra-uterine sites in which mouse blastocysts have been shown to develop. (From A. McLaren. In *The Early Conceptus, Normal and Abnormal*. Ed. W. W. Park. Univ. St Andrews Press (1965).)

females, non-pregnant females and males all give an equally high success rate. This is in strong contrast to the situation inside the uterus where, as we have seen, the hormonal conditions are so stringent that there is only a brief period in normal pregnancy when blastocysts will survive.

Another contrast between development inside and outside the uterus concerns the behaviour of the trophoblast. In extra-uterine sites, the trophoblast is more invasive, and invasive for longer. In a kidney, for example, the trophoblastic giant cells penetrate deeply into the renal tissue, engulfing the uriniferous tubules by phagocytosis and producing a haemorrhagic vesicle which may be considerably larger than the original kidney. David Kirby in Oxford showed that an important factor controlling the invasiveness of the trophoblast in the mouse uterus is the decidual reaction. When trophoblast tissue was transferred to a decidualized uterus, or to an oestrogen-sensitized uterus in which it could induce a decidual reaction, trophoblast invasion was confined to the endometrium and was no greater than in normal pregnancy. But when trophoblast tissue was introduced into a uterus incapable of mounting a decidual reaction, it not only eroded the entire endometrium, but invaded the myo-

38

metrium also, even perforating the wall of the uterus and causing substantial haemorrhage into the abdominal cavity. In
some ways the uncontrolled growth resembled that of a tumour;
it differed, however, from choriocarcinoma, the highly malignant human trophoblastic tumour, in that growth was confined
to the uterus and no secondary growths (metastases) occurred.

IMMUNOLOGICAL PROBLEMS

A skin graft, a kidney transplant, or any other tissue from a
genetically distinct individual, is briskly rejected in most
vertebrates, and certainly in all mammals. The embryo, which
is genetically distinct from the mother except in the very special
case of highly inbred lines of laboratory animals, is in intimate
contact with maternal tissues, yet it is not rejected. How does
each normal pregnancy manage to achieve for the embryo the
immunological privilege which transplant surgeons so often seek
in vain?

A limb taken from a rat fetus *in utero* and grafted to any other
part of the mother's body is at once rejected. This experiment
shows both that the embryo at this stage of gestation is capable
of eliciting an immune reaction, and that the pregnant female
is capable of mounting such a reaction. There is some evidence
that skin grafts survive a little longer on pregnant than on non-
pregnant rabbits, perhaps because of the high level of circulating corticosteroids, but this effect cannot play more than a very
minor role in the non-rejection of the embryo.

The next line of defence that might be suspected is a physical
barrier between the mother and her embryo. It is the trophoblast
rather than the embryo itself that is in contact with maternal
tissue, and in some species (man, chinchilla) giant trophoblast
cells actually become detached from the placenta and circulate
in the mother's blood stream in large numbers, often ending up
lodged in the narrow capillaries of the lung (Fig. 1-9). Micro-
scopical studies have shown that the trophoblast after implantation becomes surrounded by a more or less continuous layer of

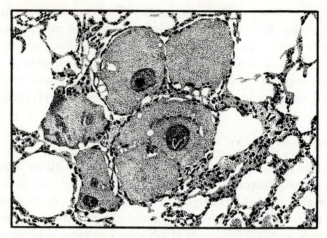

Fig. 1-9. Giant trophoblast cells of the chinchilla blocking the capillaries of the mother's lung in late pregnancy (× 120). (From W. D. Billington and B. J. Weir. *J. Reprod. Fert.* **13**, 593 (1967).)

an acidic mucopolysaccharide substance, the fibrinoid layer. David Kirby and his colleagues were the first to stress the potential importance of the fibrinoid layer in reducing immunological interaction during pregnancy. A recent experimenter has claimed that mouse trophoblast injected into an animal of a different strain provokes no state of immunity, and hence no accelerated rejection of a subsequent skin graft if the fibrinoid layer is left intact, but trophoblast with the fibrinoid layer dissolved by the enzyme neuraminidase is as powerful an immunizing agent as spleen cells or any other tissue.

However, even though the trophoblast in its fibrinoid investment may substantially reduce the embryo's immunological problems, it cannot solve them. Embryonic blood cells have been detected in the maternal circulation in both mouse and man, though reports that maternal cells normally pass the placental barrier and enter the fetal circulation have not been confirmed. In mice, both circulating and cell-bound antibodies against fetal antigens have been demonstrated in the mother. If antibodies are formed, how is the embryo protected against them?

Before implantation, there is some evidence that the zona pellucida may protect the embryo from any antibody present in the reproductive tract. During and after implantation, the uterus appears to constitute an immunologically privileged site, perhaps because of the dense layer of decidual tissue which develops in some species at this time. If mice are strongly immunized against tissue from a different strain, blastocysts from that strain will implant and grow successfully if transferred to the uterus, but show no growth if transferred to an extra-uterine site such as the kidney, where there is no protective layer of decidual tissue. Trophoblast taken from a slightly older embryo, when the fibrinoid layer has developed, will grow successfully in either kidney or uterus, irrespective of the state of immunity of the recipient.

If a mouse of one strain has borne several successive litters to a male of a different strain, the survival of skin grafts from the paternal strain is actually prolonged. In other words, immunological unresponsiveness to the paternal and hence also fetal antigens has developed. The mechanism of this unresponsiveness, which may turn out to be the most crucial single factor in preventing rejection of the fetus, is at present under investigation. It is thought to involve coating of the target tissue (skin graft or fetus) with a protective layer of harmless antibody, which blocks further immunological attack. An analogous phenomenon was demonstrated several years ago, when the presence of blocking antibodies was shown to enhance the growth of transplanted tumours. Such antibodies, often termed 'enhancing' antibodies, may play an important part in the development of immunological unresponsiveness during pregnancy.

From fertilization, through the whole pre-implantation period of development, the mammalian embryo remains relatively independent of the mother. As we shall see yet more vividly in Chapter 4, this fortunate circumstance provides experimental embryologists with a great opportunity. But from implantation

The embryo

onwards, the embryo is almost entirely at the mercy of the mother, and perishes swiftly if her hormonal, physiological and immunological adaptations to pregnancy are inadequate.

SUGGESTED FURTHER READING

The Mammalian Egg. C. R. Austin. Oxford; Blackwell Scientific Publications (1961).
Protein synthesis and enzyme constitution. R. L. Brinster. In *Regulation of Mammalian Reproduction*. National Institutes of Health (1973).
The biology of decidual cells. C. Finn. *Advances in Reproductive Physiology* **5** (1971).
Nucleic acid metabolism during early mammalian development. C. F. Graham. In *Regulation of Mammalian Reproduction*. National Institutes of Health (1973).
The design of the mouse blastocyst. C. F. Graham. In *Control Mechanisms of Growth and Differentiation*. Symposium of the Society of Experimental Biology, 25. London; Cambridge University Press (1971).
Recent studies on developmental regulation in vertebrates. A. McLaren. In *Handbook of Molecular Cytology*. Ed. A. Lima-de-Faria. Amsterdam; North-Holland (1969).
Stimulus and response during early pregnancy in the mouse. A. McLaren. *Nature, London* **221,** 739 (1969).
Blastocyst activation. A. McLaren. In *Regulation of Mammalian Reproduction*. National Institutes of Health (1973). .
Schering Symposium on Intrinsic and Extrinsic Factors in Early Mammalian Development. Venice, 1970. Ed. G. Raspé. Vieweg, Pergamon Press (1971).
The Body. A. Smith. London; Allen and Unwin (1970).
Preimplantation Stages of Pregnancy. Ed. G. E. W. Wolstenholme and M. O'Connor. Ciba Foundation Symposium. London; Churchill (1965).

2 Sex determination and differentiation
R. V. Short

Sex ought to be an easy topic to discuss; everybody is curious about and conscious of their own sexuality, and hence has at least some understanding of one half of the subject. But it would be a mistake to imagine that sex comes in only two sizes, male and female. Nor is sex determination a single, once-and-for-all decision, but rather a gradual awakening that spreads from the ovary or testis to the reproductive tract, to the general body tissues, and even to the brain itself; finally, almost the whole body is overtaken by this process of sexual differentiation.

Such a complicated sequence of events is liable to go wrong at almost any point along the line, and the anomalous individual that results may well be an intersex, and exhibit some of the structural (phenotypic) characteristics of both sexes. If an intersex individual has both ovarian and testicular tissue, either as separate entities, or as a combined organ, an ovotestis, we refer to him/her as a 'true hermaphrodite'. On the other hand, an intersex with only one type of gonad is called a 'pseudo-hermaphrodite', and is ascribed a sex depending on the appearance of the gonad. Thus a 'male pseudohermaphrodite' would have to possess testes, but the other organs could all be female.

During the last decade, our understanding of the manifold processes involved in sexual development has increased enormously as a direct result of improved research techniques. By treating suspensions of dividing cells with very weak salt solutions, it is possible to spread out their chromosomes, so that we can count them. By photographing such chromosome spreads, and then cutting out the picture of each individual chromosome and matching it with its partner, we can prepare a chromosomal photomontage which we call a karyotype. The karyotype of the male mammal usually reveals the sex chromo-

43

somes as the only two unpaired ones; once the sex chromosomes have been identified in this way, we are in a position to determine the true genetic sex of an individual. The application of such techniques to spontaneously occurring cases of intersexuality in man and animals has brought us closer to an understanding of the normal processes of sex determination and differentiation.

If we are to deal with this whole subject in a logical manner, we must begin at the beginning and consider the way in which the genes determine sex, and then go on to trace through the sequence of events that leads up to complete sexuality.

GENETIC SEX

In mammals, the female is known as the homogametic sex because she has two X-chromosomes, and all her gametes, the eggs or ova, are similar to one another in possessing a single X-chromosome. The male, on the other hand, is referred to as the heterogametic sex because he has an X- and Y-chromosome, and hence two distinct populations of spermatozoa, one X-bearing and one Y-bearing. Since the X-chromosome is usually much larger than the Y-chromosome, people have attempted to separate the two populations of spermatozoa by exploiting a possible difference in mass; so far, they have been completely unsuccessful.

It is interesting that in birds, the female is the heterogametic sex, so she lays two different kinds of eggs. In fishes and amphibians, either the male or the female can be heterogametic, depending on the species. Even in mammals, there are some interesting variants of the normal XX/XY sex-determining mechanism. In species such as the mongoose, the Y-chromosome has become joined (translocated) onto one of the autosomes (the remaining chromosomes which are not normally concerned with sex determination). In some marsupials and bats, it is the X-chromosome that has become fused to an autosome.

In addition to genes concerned with sex determination, the X-chromosome in man is known to carry at least sixty other genes for characters such as colour vision, blood clotting, a blood group substance, certain diseases such as muscular dystrophy, and many other factors. In contrast to this, the only non-sexual gene known to be on the Y-chromosome controls hairy ears! Therefore when we say that a gene is sex-linked, we mean that it is located on the X-chromosome; when we say that a gene is sex-limited, we mean that it can only be expressed in one sex; for example, the genes that control ovulation rate and milk yield are sex-limited. Genes do not seem to cross over between the X- and Y-chromosomes, or between the X-chromosome and the autosomes, and so it has been proposed that all mammals are likely to have essentially the same complement of sex-linked genes.

Since the X-chromosome obviously carries such a wealth of general genetic information, in striking contrast to the Y-chromosome, one might wonder how it is that the female can carry a double dose of many vital genes, whereas the male only has a single dose. Such inequality cannot in fact be tolerated, and so female mammals seem to have developed a dosage-compensation mechanism, whereby one of the two X-chromosomes is partially or completely inactivated. This idea was first proposed by Mary Lyon in 1961, and it has come to be known as the Lyon hypothesis. At a very early stage in embryonic development, each cell of the female embryo has to make a decision about which of its two X-chromosomes to inactivate. The decision, once made, will be passed to all the progeny of that particular cell. Using radioisotopes, it is possible to show that this inactivated or heterochromatic X-chromosome doubles up its DNA later in the mitotic cycle than the other chromosomes. If there is only one X-chromosome present, as in normal males, or females with Turner's syndrome (see Fig. 2-1), a condition described more fully in the last chapter of this book, then it will remain functional in all cells; if there are more than two X-chromosomes present, as in XXX and XXXX women, still only one remains functional, and the surplus ones are inactivated. If one of the X-chromosomes should happen to be

45

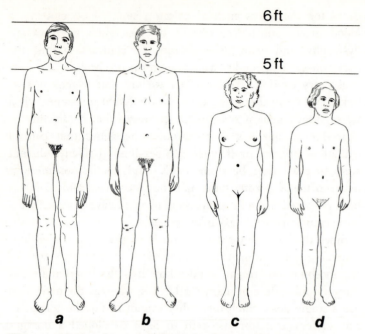

Fig. 2-1. Abnormal types of sexual development in man.
a A 24-year-old male eunuch, who had been castrated at the age of 5.
b A 30-year-old man with Klinefelter's syndrome, who is sex chromatin *positive* and has an XXY sex chromosome complement. The incidence of the condition is about 2:1000 live births.
c A 17-year-old *man* with the testicular feminization syndrome, who is sex chromatin negative and has a normal XY sex chromosome complement.
d An 18-year-old woman with Turner's syndrome, who is sex chromatin *negative* and has an XO sex chromosome complement. Many Turner's syndrome fetuses are aborted, only about 1 in 14 surviving to term, when the incidence is about 1:3000 live births.

defective, there is some evidence that it may be preferentially inactivated; otherwise, the inactivation process occurs completely at random. One can see this at work in the female tortoise-shell cat, where the coat is a mosaic patchwork of black and yellow hairs. Black hair is produced by the dominant gene, *B*, and yellow by its recessive, *b*. This gene is sex-linked, so if one X-chromosome contains the dominant gene and the other

X-chromosome the recessive, random inactivation will allow both coat colours to be expressed. Male tortoise-shell cats are understandably rare, because a normal male has no right to possess two X-chromosomes. Another interesting consequence of this random inactivation is that identical, one-egg girl twins appear to be more dissimilar in appearance than identical boys. The only cells in the body of the female where *both* X-chromosomes are apparently functional are the female primordial germ cells; this may have important consequences, as we shall see later.

The first indications of this X-inactivation process were obtained by a Canadian research worker, Murray Barr, back in 1949. He was investigating the effects of electrical stimulation of the cat's brain on the histological appearance of the nerve cells, and he noticed that a small, dense mass of heterochromatin was present in the nuclei of some cells. Then he discovered that this inclusion body was only present in female cats; so, it came to be known as sex chromatin or the Barr body (or the 'drumstick' in polymorphonuclear leucocytes), and it has since been identified in a wide variety of tissues from a number of species (see Fig. 2-2). It was really this simple discovery that sparked off renewed interest in human genetics, for it meant that anybody with a microscope, a glass slide and a bottle of stain could begin to investigate the sex chromosomes. Sex chromatin was soon put to work to predict the sex of unborn babies by examining the fetal cells that are shed into the amniotic fluid. Early blastocysts can now be sexed, giving a more accurate estimate of the true primary sex ratio at fertilization. One can even sex blood smears by looking for the drumstick-shaped nuclear appendages of certain white cells (polymorphonuclear leucocytes), and this has been used in forensic medicine. Sex chromatin has also provided us with a simple clinical diagnostic test for sex chromosome abnormalities, because the number of sex inclusions indicates the number of inactivated X-chromosomes (see Table 2-1). If a man is 'sex chromatin positive', showing a sex inclusion, then there must be two X-chromosomes present. He is therefore likely to be a case

47

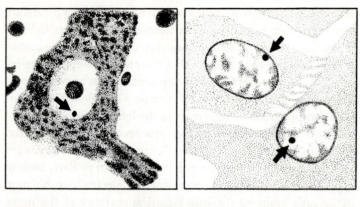

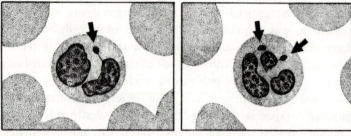

Fig. 2-2. Intranuclear sex inclusion or Barr body (arrowed) in a female (XX) nerve cell (*top left*), and in two female fibroblasts (*top right*). One 'drumstick' is seen in a normal female polymorphonuclear leucocyte (*bottom left*), and two in a polymorph from an XXX woman (*bottom right*).

TABLE 2-1

	Male phenotype	No. of sex inclusions	Female phenotype	No. of sex inclusions
Genetic	XY	0	XX	1
constitution	XXY	1	XO	0
	XYY	0	XXX	2
	XXYY	1	XXXX	3
	XXXY	2	XXXXX	4

of Klinefelter's syndrome, XXY (see Table 2-1). This too is discussed more fully in the last chapter.

Once the genetic balance of the early embryo is restored by making only one X-chromosome functional in the female, the stage is set for sexual differentiation. The scene now shifts to the genital ridges, which develop in the dorsal body wall of the embryo as the forerunners of the gonads (see Book 1, Chapter 1). It is only in these sites that genetic sex seems to be fully expressed; most of the sexual information coded in the nuclei of all the other tissues throughout the rest of the body probably lies dormant for the remainder of the animal's life. Certainly, it is true that many of the somatic tissues will have some degree of sexual differentiation imposed upon them at a later date, but this will be brought about by the action of sex hormones. The delegation of sexual differentiation to hormones is typically mammalian; in insects, for example, there is evidence that it is the actual genetic sex of cells in the brain that controls the animal's sexual behaviour.

Perhaps the most striking illustration of the all-powerful effects of the gonads on sexual development comes from the rare recorded cases of 'monozygotic' or one-egg human twins, one of whom was a boy and one a girl. The explanation for this incredible situation is that soon after fertilization, one cell in a developing male embryo lost a Y-chromosome. The embryo then divided into two, and as a result the tissues of both twins were a mixture (mosaic) of XY and XO cells; in one twin, XY cells predominated in the genital ridge, causing it to develop as a testis, whereas in the other twin XO cells gave rise to an ovary. Subsequent sexual development of the twins along different paths was determined by this one event.

From what we have said so far, you will already have begun to suspect that it is the Y-chromosome that holds the key to sex determination, and this view receives considerable support if we look at the appearance (phenotype) of people with a variety of sex-chromosome abnormalities (see Table 2-1).

The absence of *any* X-chromosomes is incompatible with

survival; if a Y-chromosome is present, the individual will be a male, regardless of the number of X-chromosomes he possesses. But in the *absence* of a Y-chromosome, does one always get a female? Until a few years ago, this was thought to be the case. Then a number of individuals were discovered who had a male phenotype, but an XX-chromosomal constitution, and no sign of a Y-chromosome anywhere. They were initially dismissed as undiagnosed mosaics, who must have XY-cells hidden somewhere in the body. But studies now going on in goats, pigs and mice have forced us to change our views; in each of these species, there appear to be autosomal genes that can sometimes cause genetic females (XX) to turn into phenotypic males with well-developed testes and male reproductive tracts. At least in the mouse and goat this cannot be explained by translocation of the Y-chromosome onto an autosome, and so it has been suggested that the male-determining genes themselves may actually be located on the X-chromosome, and that the Y-chromosome, or an autosomal mimic of it, merely acts by 'switching on' these X-linked male genes.

Thus far, we have only considered the action of the sex chromosomes in the somatic cells of the genital ridge. But the gonad is a consortium of somatic and germinal tissues (Book 1, Chapter 1), and there is evidence to suggest that the sex chromosomes are also at work in the germ cells, although in a somewhat different way. We have already remarked on the fact that *both* X-chromosomes remain active in female germ cells. In Turner's syndrome (XO) in women, the primordial germ cells start to develop normally, but they have nearly all perished by the time of birth, suggesting that in man the second X is necessary for germ cell survival; paradoxically, XO mice appear to be perfectly fertile.

If XX primordial germ cells are placed in a testicular environment, as can happen in experimental 'chimaeras' (described in the first and fourth chapters of this book), or in the sex-reversed XX goats, pigs or mice, then they are unable to survive. Similarly, the XXY primordial germ cells in males with Klinefelter's syndrome cannot survive in the testis. Thus there are important interactions between the genetic constitution of these primordial germ cells and their environment. The mere presence of a testis cannot induce the germinal cells that migrate into it to undergo spermatogenesis. Unfortunately, we do not know what would happen to XY primordial germ cells in an ovary.

Whilst the number of primordial germ cells in no way influences the endocrine activity of the testis, the situation in the ovary is very different. It is the female germ cells that induce the development of the follicle cells, and these in turn give rise to the Graafian follicles which are the future endocrine apparatus of the ovary; before ovulation, the follicle is the site of oestrogen synthesis, and after ovulation it gives rise to the progesterone-secreting corpus luteum (see the first and third chapters in Book 3). And so a genetic defect in the female germ cells, as in Turner's syndrome in women, can result in a gonad with no endocrine activity, and hence a failure of all female secondary sexual characteristics. Germ-cell sex therefore controls both the fertility and the endocrine activity of the ovary, but the testis can continue to secrete testosterone even if there are no germ cells present at all.

GONADAL SEX

The way in which the primordia of the gonads can differentiate into either a testis or an ovary has already been discussed in the first chapter of Book 1. Somehow, male-determining genes cause the persistence of a centrally placed mass of cells, the so-called medullary sex cords, which eventually give rise to the seminiferous tubules of the testis; female-determining genes, or maybe simply the absence of male-determining ones, result in regression of this medullary tissue and an ingrowth of cortical tissue from the surface layers of the gonad to form the substance of the ovary. Experiments in which male and female amphibian larvae have been surgically united to one another so that their circulations fused, gave rise to the concept of 'cortico-medullary

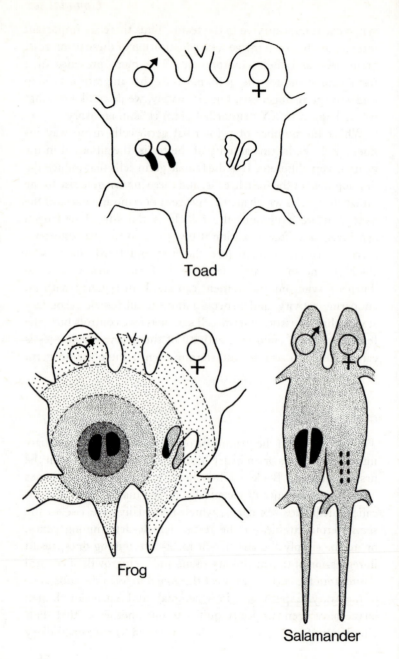

Toad

Frog

Salamander

antagonism' (see Fig. 2-3). If the medulla was stimulated to develop, it was thought to produce a substance that could act locally or via the circulation to inhibit the ingrowth of ovarian cortex, and vice versa.

The male-determining genes themselves could act as simple gonadal growth promoters, since during embryonic development testicular growth is considerably in advance of ovarian growth. Experiments in which rat fetal ovaries and testes have been transplanted close to one another beneath the capsule of the kidney in adult rats have provided evidence in favour of some sort of cortico-medullary antagonism in mammals, and some circulating inductor substance could explain the gonadal sex reversal that occurs in 'freemartins'. In this condition, which occurs in cows, sheep, pigs and goats, a female fetus that is twin to a male and shares a conjoined placental circulation (see Fig. 2-4) undergoes a partial sex reversal of her ovaries, which interferes with the subsequent sexual differentiation of the female reproductive tract (see Fig. 2-5). It used to be thought that freemartins were produced by the action of male sex hormone (androgen) from the testis of the bull calf passing into the female and masculinizing the gonads and reproductive tract. But we now know that androgens cannot turn a mammalian ovary into a testis, nor can they cause the regression of the uterus that is such a typical feature of all freemartins.

Tempting though the idea is of cortico-medullary antagonism, before we can accept it we must be able to explain those cases of true hermaphroditism where one or both gonads are ovotestes.

Fig. 2-3. Experimental evidence for a gonadal inductor substance. Male and female toads, frogs and salamanders have been surgically united to one another before metamorphosis, so that they share a common circulation. In the case of the toad, the testes and ovaries of the male and female larvae continue to develop normally. In the frog, however, some substance manufactured by the testis passes across into the female and partially interferes with the development of the ovaries. In the salamander, the male gonadal inductor completely inhibits ovarian development (From E. Witschi. *Development of Vertebrates*, W. B. Saunders Company (1956).)

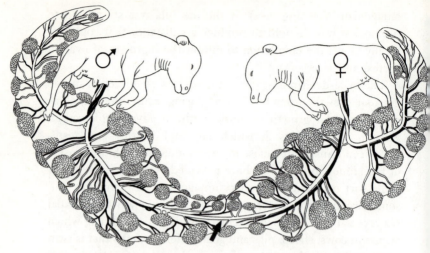

Fig. 2-4. The cause of the freemartin condition. When twin fetuses of opposite sex share a common circulation as a result of fusion of placental blood vessels (arrowed), some substance passes from the male to the female, causing partial sex-reversal of her ovaries. This interferes with the normal development of the female reproductive tract, so that the animal is sterile. (From F. R. Lillie. *J. Exp. Zool.* **23**, 271 (1917).)

This condition is particularly common in sows, and the odd thing is that they may be fertile if one gonad is an ovary, regardless of the amount of testicular tissue in the other gonad.

It is curious that one can produce complete, functional sex reversals in fish and amphibian larvae by exposing males to oestrogens or females to androgens. Furthermore, these sex-reversed individuals may be fully fertile. In mammals, however, this is never possible and neither androgen nor oestrogen treatment of the mammalian fetus can cause an ovary to change into a testis or vice versa. This loss of gonadal plasticity in mammals may have been an essential requirement for the development of viviparity; if the fetus has to develop within its mother's abdomen, it will be subjected to female sex hormones manufactured by the maternal ovaries and also maybe by its own placenta. The gonads of the male fetus therefore need to be protected against this feminizing influence.

Of one thing we can be quite certain; if medullary cords persist in the gonad they will secrete androgens, irrespective of their genetic sex. So hormone production is controlled by the structure of the gonadal tissue rather than by its chromosomal constitution. This brings us to our next topic.

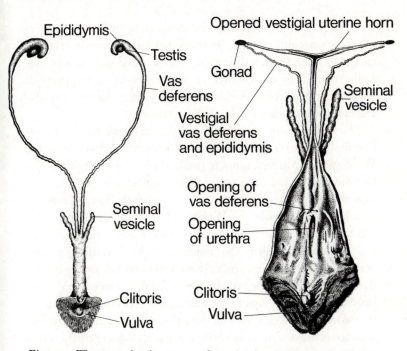

Fig. 2-5. The reproductive tracts of two bovine freemartins, showing the great variability that exists. On the left is the tract of a newborn animal described by Lillie with complete masculinization of the internal genitalia. The gonads, although very small, have been completely transformed into testes and have descended into the inguinal canals. The epididymi, vasa deferentia and seminal vesicles are well developed, and all remnants of the Mullerian duct have been completely suppressed. However, the external genitalia are still female in appearance, apart from slight hypertrophy of the clitoris. On the right is the reproductive tract of an adult freemartin, dissected and drawn for John Hunter 200 years ago. The gonads are extremely small, and both male and female duct systems are present, but incompletely developed. The external genitalia are those of a normal cow, although the clitoris is slightly hypertrophied.

55

Sex determination and differentiation

What hormones are manufactured by the fetal gonads, and what do they do? The answer to part of this question can be given in the form of an equation:

$$\male \neq \text{\male\kern-0.8em} = \female = \female.$$

In other words, the testis of the developing male fetus super-imposes masculinity on a basically neutral or female state. A castrated male fetus will develop in the same way as a female fetus, regardless of whether or not her ovaries are present, emphasizing not only the dominant role of maleness, but also the completely passive role of the fetal ovary.

The embryological primordia of the male and female reproductive tract are two duct systems, known respectively as the Wolffian and Mullerian ducts (males are wolffs, for those who like mnemonics). In the male, each Wolffian duct will develop into an epididymis, vas deferens, ampulla of the vas and seminal vesicle. In the female the Wolffian duct never develops at all, but the Mullerian duct, whose development is inhibited in the male, persists to form the Fallopian tube (oviduct), uterus, cervix and part of the vagina; these male and female organs are described briefly in the last chapter of Book 1. Since the Mullerian duct never develops in the intact male fetus but is present in the castrated male, it seems obvious that some hormone from the fetal testis must *actively inhibit* the Mullerian duct; by the same token, the fetal testis must *actively stimulate* the Wolffian duct.

Studies on castrated male fetuses treated with androgen, and more recently the use of a synthetic anti-androgenic steroid, cyproterone, which specifically blocks the actions of androgen on its target organs, have shown that the Wolffian-stimulating activity of the fetal testis is androgen-dependent, but the Mullerian-inhibiting activity is *not* brought about by androgens. The fetal testis must therefore produce a *second* hormone, but at the moment we have no clues as to what it might be.

It is interesting to consider for a moment the different effects produced by castration of a male fetus followed by testosterone treatment, or anti-androgen administration to an intact male fetus. Following castration + testosterone, the Wolffian ducts will develop into a normal male reproductive tract; but removal of the testis will have eliminated the source of Mullerian-inhibiting hormone, so we get an animal with fully developed male *and* female reproductive tracts. In contrast, when anti-androgen is given to an intact male fetus, the Wolffian ducts fail to develop; however, the testis continues to produce the Mullerian-inhibiting hormone, so the Mullerian ducts also fail to develop. We therefore have an animal with testes, but no reproductive tract at all, as in the testicular feminization syndrome (see Fig. 2-1 and next section).

One final point of interest concerns the site of action of these two fetal testicular hormones. The classical experiments of Alfred Jost in Paris showed that if a piece of testis was grafted onto one of the ovaries of a female rabbit fetus, then the reproductive tract on that side of the body would be masculin-ized, whereas the other side would be unaffected (see Fig. 2-6). This emphasizes the fact that the fetal testis exerts its greatest effect on the structures closest to it; it seems that inhibition of the Fallopian tubes and uterus, and stimulation of the epididymis and vas deferens, are *local* events, not under the control of circulating hormones. The masculinization of the external genitalia, on the other hand, is produced by circulating andro-gens, and an increase in the ano-genital (anus to phallus) distance is one of the most sensitive indices of virilization of a female fetus that has been under the influence of androgens (see Figs. 2-7 and 2-8).

Translated into practical terms, these observations explain why one finds a bilaterally asymmetrical reproductive tract in true hermaphrodites where one gonad is a testis and one an ovary; on the side of the testis, there will be a male duct system, and a female duct on the ovarian side. This is not an uncom-mon finding in intersex pigs, and any animal showing this type of asymmetry is technically referred to as a gynandromorph.

57

Sex determination and differentiation

The distinction between the local and generalized actions of the fetal testis are also well illustrated in the case of freemartins. Following the partial conversion of the fetal ovary into a testis, it is not uncommon to find complete masculinization of the internal duct systems down to the level of the seminal vesicles, whilst the external genitalia remain essentially female, with perhaps some slight hypertrophy of the clitoris (see Fig. 2-5).

PHENOTYPIC SEX

You might imagine that your sexual appearance is an automatic consequence of your hormonal sex; once an androgen-produc-

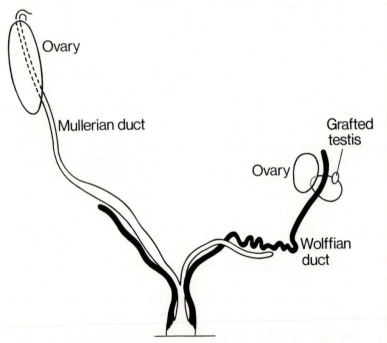

Fig. 2-6. The classical experiment of Alfred Jost, where he grafted a piece of testis against the right ovary of a rabbit fetus. The testis produced a local stimulation of the Wolffian duct and inhibition of the Mullerian duct, but the effect did not extend to the other side of the body.

ing testis, always a male. This is indeed generally true, but we
have recently come to realize that there is an important exception, which is seen in the testicular feminization syndrome.
This has been described in human beings, cows, sheep, rats and
mice (see Fig. 2-1). Individuals that are genetic males (XY)

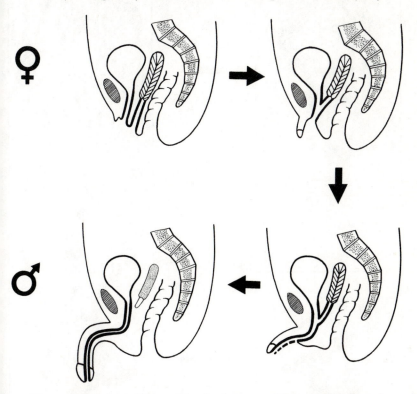

Fig. 2-7. Diagram to illustrate the development of the normal male and
female reproductive tract in man, and varying degrees of intersexuality.
In the normal female (*top left*), the clitoris is vestigial and the urinary
and genital ducts open separately. This basic pattern is established by
the 16th week of gestation, and cannot subsequently be altered.
Masculinizing influences before the 16th week can produce enlargement of the clitoris and fusion of the urinary and genital ducts (*top
right*), or an actual phallus with a penile urethra which may open at its
base (*bottom right*). In the normal male (*bottom left*) the uterus has
become vestigial. (From C. Overzier. *Intersexuality*. Academic Press
(1963). Permission of Georg Thieme Verlag, Stuttgart.)

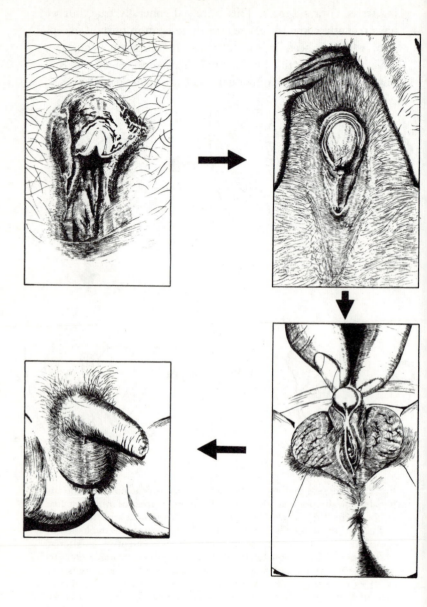

develop testes in the normal way, and the testes secrete normal amounts of androgen and Mullerian duct inhibitor. But because of a genetic defect, none of the target organs are able to respond to the circulating androgen, so that no male duct system develops. This defect does not interfere with the action of the Mullerian duct inhibitor, so no female duct system develops either. Externally, the resultant individuals are phenotypic females, and affected 'women' have good breast development. However, they do not menstruate and usually lack pubic hair; injections of androgen are unable to produce any signs of masculinization, which is hardly surprising since they already have normal male levels of testosterone in the peripheral blood.

In mice, we now know that the gene responsible for testicular feminization is located on the X-chromosome, and the same is probably true of man and the domestic animals. As yet, we do not understand the nature of this genetic defect; at one stage it seemed as if it might be due to a lack of an enzyme in target organs that normally converts testosterone into its androgenically active breakdown product, dihydrotestosterone, but this theory has now been largely discounted.

A more detailed study of testicular feminization may well lead to a closer understanding of how it is that testosterone normally brings about its effects in target tissues. It also provides us with an interesting exception to the general rule that the sex

---

Fig. 2-8. Various degrees of intersexuality encountered in girls suffering from congenital adrenal hyperplasia (the adrenogenital syndrome). In the mild form (*top left*) there may be only a slight enlargement of the clitoris, and a premature development of pubic hair. In a more extreme case (*top right*) the clitoris may enlarge still further, and begin to develop a prepuce. The urethra and vagina now open into a common urogenital canal. If masculinization is even more extensive (*bottom right*), the urogenital canal may open at the root of the penis, and the lips of the vulva on either side of the clitoris begin to form into a bipartite scrotum. Finally (*bottom left*) the genitalia may become completely masculinized, except that the scrotum will be empty, and normal ovaries will be found in the abdomen, together with a uterus. (From C. Overzier. *Intersexuality*. Academic Press (1963). Permission of Georg Thieme Verlag, Stuttgart.)

chromosomes of somatic cells have no part to play in their sexual differentiation. Testicular feminization patients pose some ethical problems for the clinician, too. They are often voluptuously beautiful women, who get married and then go to an infertility clinic because they are not menstruating. On clinical examination, the vagina is found to be blind-ending, and testes can often be palpated in the groin. Because of the very real danger that these intra-abdominal testes will develop into malignant tumours, it is necessary to remove them. The operation cannot be performed without the patient's consent, and if she came to the doctor in the first place for advice about an infertility problem, she is not likely to consent to removal of her gonads. Yet it would be a great psychological shock for a woman to learn that she was in fact a man with testes that might become cancerous.

Phenotypic sex may also be at variance with gonadal sex in human babies suffering from overdeveloped adrenal glands (congenital adrenal hyperplasia, or the adrenogenital syndrome). This condition is genetically determined, and results from the loss of one or other of the enzymes in the adrenal cortex that are responsible for synthesizing the principal adrenal hormones like aldosterone, corticosterone and cortisol. The activity of the adrenal cortex is normally controlled by the secretion of adrenocorticotrophic hormone (ACTH) from the anterior pituitary, and this in turn is regulated by a negative feedback of cortisol on the hypothalamus (these mechanisms are described more fully in the first chapter of Book 3). If the adrenal cannot produce adequate amounts of cortisol because of the enzyme defect, then the pituitary will attempt to compensate by secreting an increased amount of ACTH, which in turn causes excessive enlargement of the fetal adrenals. Now the adrenal gland normally secretes only trace amounts of sex hormones, particularly androgens. But in congenital adrenal hyperplasia, this androgen secretion may reach significant proportions, and can cause extensive masculinization of the external genitalia of baby girls (see Fig. 2-8). The clitoris enlarges into a penis-like structure, and the labiae of the vulva may fuse to give a type of scrotum. Plastic surgery is required to restore these infants to their true phenotypic sex, and treatment with adrenal corticoids to make up for the cortisol deficiency. Baby boys suffering from congenital adrenal hyperplasia may show the 'Infant Hercules' syndrome, and reach puberty when only a few months or years old; but fortunately, they do not require plastic surgery.

Similar masculinization changes in newborn infants are sometimes seen if a pregnant mother has been on a course of treatment with androgens. Some of the early synthetic sex hormones that were used for the treatment of threatened or habitual abortion unfortunately produced androgenic side-effects, and the consequences were distressing if the mother was carrying a female fetus.

The popular press and the general public have a morbid fascination for those unfortunate individuals who desire to 'change' their sex by altering their appearance. If this merely takes the form of wearing the clothes of the opposite sex, the individual is known as a 'transvestite'; it is commoner for men to desire to dress as women than vice versa. In more extreme cases, the individual may actually alter his or her physical appearance, and so would be described as a 'transsexual'. If a man puts himself onto continuous oestrogen therapy, he will be able to induce a considerable degree of breast development. Repeated manual stimulation of the nipples, or prolonged administration of large doses of tranquillizers, may even induce lactation in men. In extreme cases, male transsexuals may submit to castration, and amputation of the penis. An artificial vagina is created by making an opening in front of the anus, and lining it by turning in the skin of the penis; part of the scrotum can then be used to emulate the female labiae. Casablanca seems to have become the centre for this type of operation, and following this anatomical and psychological mutilation the transsexual is capable of having intercourse with a man.

If a woman is put onto long-term androgen treatment for medical or psychological reasons, she will begin to develop a male-type distribution of facial and body hair, and may even show incipient signs of baldness and a change in voice. However, there will be little effect on the external genitalia other than some slight enlargement of the clitoris. The genital primordia are extremely sensitive to androgens during embryonic development, but once the adult pattern has become established, little further change is possible. Once a male pattern of hair growth has developed in a man or woman, this is also extremely difficult to change, even if one resorts to castration and female sex-hormone therapy.

Spontaneous changes in phenotypic sex can occur in response to hormone-secreting tumours of endocrine organs. Perhaps the commonest example is seen in the case of dogs with tumours of the testis; these animals may come into 'oestrus', attracting other male dogs from miles around, and the mammary glands may develop to the point of lactation. Alterations in adrenal or ovarian androgen secretion in women can give rise to male patterns of hair growth which are psychologically most distressing.

Some outstanding examples of spontaneous sex reversal are seen in birds. In some species of birds only one of the two gonads normally develops in the female to form a single ovary. If this is removed surgically, or damaged by disease, the gonadal rudiment on the other side may enlarge and develop into an androgen-secreting testis. In chickens, the comb begins to grow to the size of a cockerel's, and the bird begins to crow. The plumage may also change to that of the male, and this can be particularly spectacular in the case of pheasants. There are even rare cases on record of an egg-laying chicken undergoing a complete, functional sex reversal to become a sperm-producing rooster.

It is interesting to contrast the situation in birds to that in mammals. In both cases, it is the heterogametic sex (XY) that is dominant to the homogametic sex (XX). But in birds, the heterogametic sex exerts its dominance by means of an oestrogen-secreting ovary, as opposed to the androgen-secreting testis of the mammal.

We might sum up the attitude of society to cases of 'sex change' by the old adage:

'A whistling woman and a crowing hen
Are neither good to God nor men.'

BRAIN SEX

The testis and the ovary are known to be controlled by the same pituitary hormones. However, if ovaries are transplanted into castrated males, they cease to show the cyclical patterns of follicular growth and ovulation that are such a characteristic feature of the normal ovary. This naturally led people to suspect that there might be a sex difference in the pituitary gland itself; but the idea was soon disproved by surgical transplantation of pituitaries from male rats into female rats whose own pituitaries had previously been removed. These female rats with male pituitaries subsequently showed normal, regular oestrous cycles. When people began to realize that the pituitary itself was under the control of another area of the brain, the hypothalamus, attention was switched to the idea of 'hypothalamic sex'. It was shown that a single injection of androgen given to a newborn female rat, mouse or hamster could effectively sterilize it for the rest of its life by producing a state of persistent oestrus. It has been suggested that male animals, or females given an injection of androgen within a day or two of birth, develop a 'continuous' pattern of hypothalamic activity which results in a continuous release of gonadotrophic (gonad-stimulating) hormones from the pituitary. Female rats, or castrated newborn males, or even newborn males treated with the anti-androgen, cyproterone, develop a 'cyclical' pattern of hypothalamic and pituitary activity which is responsible for the cyclical recurrence of oestrus in the normal female. We still do not have any direct proof of these two differing patterns of hypothalamic activity, and the experimental results could equally well be explained by the differing sensitivities of male and female hypothalami to the feedback effects of gonadal hormones; this whole subject is discussed in more detail in Book 3, Chapter 3 and Book 4, Chapter 2. However, we do now have biochemical evidence to show that the testis of the newborn rat produces a sudden surge in testosterone secretion which is sufficient to imprint the hypothalamus with the male pattern of activity. Paradoxically, small doses of oestrogen will also masculinize the hypothalamus. As it was with the development of the gonads themselves and the genital ducts, so it is with the brain. The male pattern is superimposed on the neutral or female state.

Rats, mice and hamsters are very immature at birth, whereas guinea pigs are well developed. Therefore it is not surprising that the critical period for hypothalamic imprinting in the guinea pig appears to be halfway through gestation, instead of immediately after birth. In Rhesus monkeys, androgens have been given to pregnant females from the earliest stages of gestation, and even though the female fetuses have shown extensive masculinization of their external genitalia at birth and have subsequently been slow to attain puberty, they have eventually had normal, ovulatory menstrual cycles. Clinical evidence in man also suggests that baby girls, although born with extensive masculinization of the genitalia from congenital adrenal hyperplasia, may nevertheless menstruate regularly after puberty.

Clearly, we still have much to learn about brain sex. But there can be no doubt about its existence, at least in rodents. Brain sex determines the pattern of gonadal activity, but how brain sex may go on to influence the animal's behavioural responses to sex hormones is another topic, which we must discuss in the next section.

BEHAVIOURAL SEX

We normally think of sexual behaviour as being an immediate consequence of sex hormone secretion. In most female mammals, oestrogen secretion rises suddenly just before ovulation, coupled with an increase in androgen secretion. Progesterone secretion may also be changing at this time, and it is this complex sequence of events that brings about the phenomenon of behavioural oestrus (see Chapter 3 in Book 3 and Chapter 2 in Book 4 for further details). The word 'oestrus' comes from the Greek for a gadfly, an insect that can drive cattle into a state of frenzy; the word is therefore used to convey some idea of the hyper-excitable state of the female at this stage of the reproductive cycle: rats run around, mares kick and squeal, cows bellow and mount one another, cats call, and bitches desert their owners. But oestrus is primarily a time when the female attracts the male by visual and olfactory cues, and may actively seek his company. One of the cardinal signs of oestrus is that the female will stand still when the male attempts to mount, and will show lordosis (hollowing of the back). The only species of mammal that shows no outward anatomical or behavioural change around the time of ovulation is *Homo sapiens*, and perhaps this is just as well for the maintenance of law and order in our society.

Male sexual behaviour is less well documented, and people have tended to concentrate on copulatory behaviour itself, and its component parts such as mounting, intromission, thrusting and ejaculation. However, the study of a seasonally breeding male with a well-defined period of rut, such as the Red deer stag, shows that there is in fact a whole spectrum of testosterone-induced behavioural changes that precedes the onset of copulatory behaviour itself.

We can draw a broad general distinction between male and female sexual desire, or libido; female mammals show rapid responses to sex hormones, whereas in males the changes are slow and insidious, and may be markedly influenced by prior sexual experience. Removal of the ovaries causes an immediate cessation of oestrous behaviour, and oestrogen injections will induce oestrus within 1 or 2 days. But if an adult ram is castrated, it may be a year or so before he ceases to mount oestrous ewes, and if men are castrated as adults, it may be years before they lose their libido. This has given rise to the concept that, with the increasing development of the brain, there has been an increasing emancipation of sexual behaviour from sex-hormone secretion.

Male and female libido are clearly very different types of behaviour. So we must ask ourselves whether this is due to the different hormones secreted by the gonads, androgens versus oestrogens, or whether the difference is even more fundamental.

Physiological doses of oestrogens given to male rats, or androgens given to female rats, do not necessarily change the behaviour of the individual to that which is characteristic of the opposite sex; hence we must direct our enquiries more towards the way in which the brain responds to the hormone, rather than to the presence of different hormones. Are male and female libido mutually exclusive events, so that the brain can only exhibit either one or the other type of sexual response? This seemed an attractive idea at one time, but evidence is now hardening in favour of the concept of two separate 'centres' for sexual behaviour, one for male-type and one for female-type. Which of these two develops its full potential depends on the endocrine environment of the fetus and newborn; under certain experimental conditions it is possible to develop both potentialities simultaneously.

If male rats are castrated at birth, they will grow up to show a good lordosis response to oestrogens, but a poor copulatory response to testosterone. Normal male rats, on the other hand, require large amounts of oestrogen before they will show any lordosis. This is somewhat reminiscent of the factors determining phenotypic sex; it appears to be the secretion of testosterone during a certain critical period in the newborn male rat that allows the animal to show a male behavioural response to testosterone secretion after puberty, and suppresses the oestrous response to female sex hormones. In the absence of this imprinting secretion, the animal will be incapable of exhibiting male libido later in life, but will retain a normal female libido. It has even been claimed that such androgen-deprived males may show an *oestrous* response to injected androgen; this could possibly have some bearing on the causation of homosexuality in the human male.

When we come on to the factors that control human sexual behaviour, we sail into an uncharted ocean. Extensive studies of male homosexuals have usually failed to reveal any abnormality of hormone secretion, and a number of them are known to lead a normal heterosexual existence whilst still maintaining homo-sexual relationships. Human sexual behaviour is strongly in-fluenced by psychological factors, such as the assigned gender role during the first few years of life; if a boy is reared as a girl, or vice versa, this may have far-reaching psychosexual conse-quences. The endocrine environment during intra-uterine life is also critically important: a girl exposed to androgens during this time will show marked 'tomboyish' behaviour in childhood, and delayed menarche.

To those who find male or female homosexual behaviour repugnant, it is as well to remember that some degree of homosexual activity is the rule rather than the exception amongst juveniles in free-living communities of animals. And if the heterosexual practices of this permissive society seem deviant, we should remember that in the wild, the prepubertal male chimpanzee may learn his sexual behaviour by copulating with his mother.

LEGAL SEX

In contrast to the discussions in the rest of this chapter, where we have recognized many different types of sex and degrees of intersexuality, the Law is absolute and uncompromising; it recognizes only two sexes, male and female. The first instance in English law of a legal definition of sex was the case of Corbett *v.* Corbett (otherwise Ashley) in 1970. A marriage was annulled when the 'wife', a man who had undergone a sex-change operation, was adjudged to be a male. In his summing-up, the judge said: 'The law should adopt, in the first place, the first three of the doctors' criteria, i.e. the chromosomal, gonadal and genital tests, and, if all three are congruent, determine the sex for the purpose of marriage accordingly, and ignore any operative intervention'. In cases where the three criteria are not congruent, 'greater weight should probably be given to the genital criteria than to the other two'. This seems an eminently sensible judgement, because the assigned gender at birth will usually have been determined by the phenotypic sex, rather than the chromosomal or gonadal sex. It is clearly right that testicular feminization patients should be considered as female; they will have been reared as girls, and look like girls, even though their appearance does happen to be at variance with their genetic and gonadal sex (see Fig. 2-1).

In conclusion, how can we attempt to summarize all that has been said in this chapter? In all the systems that we have considered, maleness means mastery; the Y-chromosome over the X, the medulla over the cortex, androgen over oestrogen. So physiologically speaking, there is no justification for believing in the equality of the sexes; *vive la différence*!

SUGGESTED FURTHER READING

Role of hormones in the differentiation of sex. R. K. Burns. In *Sex and Internal Secretions*, 3rd edition, vol. 1, p. 76. Baltimore; Wilkins Co. (1961).

Experimental control of psychosexuality. R. W. Goy, *Philosophical Transactions of the Royal Society of London*, B **259**, 149 (1970).

Significance of sex chromosome derived heterochromatin in mammals. J. L. Hamerton. *Nature, London* **219**, 910 (1968).

Hormonal differentiation of the developing central nervous system with respect to patterns of endocrine function. G. W. Harris. *Philosophical Transactions of the Royal Society of London*, B **259**, 165 (1970).

Modalities in the action of gonadal and gonad-stimulating hormones in the foetus. A. Jost. *Memoirs Society for Endocrinology* **4**, 237 (1955).

Genetic activity of sex chromosomes in somatic cells of mammals. M. F. Lyon. *Philosophical Transactions of the Royal Society of London*, B **259**, 41 (1970).

The bovine freemartin: a new look at an old problem. R. V. Short. *Philosophical Transactions of the Royal Society of London*, B **259**, 141 (1970).

Suggested further reading

Sex chromosomes. U. Mittwoch. London; Academic Press (1967).
Sex chromosomes and sex-linked genes. S. Ohno. Berlin; Springer-Verlag (1967).
Intersexuality. C. Overzier. London; Academic Press (1963).
Intersexuality. Ed. J. S. Perry, *Journal of Reproduction and Fertility,* Supplement 7 (1969).
Man and Woman, Boy and Girl. J. Money and A. A. Ehrhardt. Baltimore; Johns Hopkins (1972).
Germ cell sex. R. V. Short. In *International Symposium on the Genetics of the Spermatozoon,* pp. 325-45. Edinburgh (1972).

3 The fetus and birth
G. C. Liggins

Cut off from inquisitive eyes, as the fetus is, by the body wall of the mother and by the wall of the uterus, it is all too easily imagined as a sort of hothouse plant doing little more than grow during the weeks or months *in utero*, but ready to burst into full flower at birth. Even the greatest of fetal physiologists, Sir Joseph Barcroft, born 100 years ago, thought that the fetus lived in a sensory void, oblivious of its environment. But a moment's reflection is sufficient to reveal that birth represents no more than a transition from one environment to another and that most of the functions and activities to be observed in a newborn must have been present before birth. New techniques allowing study of the fetus in an undisturbed environment have convincingly confirmed that this is so. As a consequence there has been a tendency to replace the 'hothouse plant' concept by one that sees the fetus as a miniature adult, indulging in many of the activities of its extra-uterine counterpart. This concept is little better, for on the one hand it minimizes the immaturity of many important organ systems at birth and on the other hand it ignores the many specialized functions peculiar to the fetus, upon which successful pregnancy depends. As this chapter proceeds, a picture will emerge of the fetus as an individual, dependent on the mother for the ingredients of growth and survival but nevertheless enjoying a remarkable degree of independence in the regulation of its development. And furthermore, by means of hormones secreted into the maternal circulation by the placenta or by the corpus luteum, the fetus can exert effects on the maternal physiology that enhance the mother's ability to meet the metabolic requirements of pregnancy. Finally, we shall see that not only may the fetus determine the timing of birth, but also, in anticipation of this event, initiate lactation.

FETAL GROWTH

A fetus does not differ from a young animal in needing an adequate supply of nutrient material to maintain normal growth. But unlike the young animal whose diet is a complex mixture of carbohydrates, fats and proteins, the fetus has a diet consisting largely of glucose, together with sufficient quantities of amino acids to satisfy the nitrogen requirements of protein synthesis, and with small amounts of materials such as fatty acids, vitamins, salts and metals that are essential to normal growth and function. Glucose is the exclusive source of the abundant energy needed for the synthesis of tissues and, therefore, to know where the glucose comes from, how it gets to the fetus and how the fetus regulates its use, is important.

The source of glucose. The immediate source of glucose for the fetus is the maternal blood stream. Indirectly, glucose comes from food consumed by the mother, from stores of glycogen, particularly in the liver, from fat depots and, during periods of starvation, from the breakdown of protein. The amount of glucose available to the fetus depends on its concentration in the maternal blood stream which is maintained within rather narrow limits by a complex control system involving several endocrine organs. On the one hand, maternal glucose levels are prevented from falling by absorption of glucose from the gut and by the action of growth hormone, corticosteroids, catecholamines and glucagon which increase liberation of glucose from body stores. On the other hand, an increase in glucose concentration above normal limits is prevented by insulin, which increases both the breakdown of glucose in muscles and its passage into glycogen or fat stores. This system serves the adult well and ensures adequate supplies of glucose in a wide variety of stressful situations. Yet these safeguards do not always seem sufficient to satisfy the demands of reproduction, and certain pregnancy hormones may further modify maternal glucose control in a way that increases availability of glucose to the fetus. In many species, during the first half or more of

73

pregnancy, there is a considerable increase in body weight due to the deposition of fat. This is an effect of progesterone which not only increases appetite but also diverts glucose into fat synthesis. Later in pregnancy when the metabolic requirements of the fetus are at their peak, or during periods of starvation, these extra fat stores are used to maintain a constant supply of glucose to the fetus. Pregnant primates have an additional hormone with the tongue-teasing title of human chorionic somatomammotrophin (HCS), which is secreted by the placenta in large quantities. HCS is closely related chemically to growth hormone, and although its biological activity is low, it shares with growth hormone an effect in elevating blood glucose levels.

Placental transport of glucose. As we might expect for such an important substance, the passage of glucose from maternal to fetal circulations occurs readily by means of an active transport system. When glucose is injected into experimental animals or into pregnant women, a rise in the concentration of glucose in maternal blood is followed rapidly by a comparable increase in the concentration in fetal blood. The two levels do not become equal, for a gradient of concentration from mother to fetus is always present, the concentration in maternal blood being higher. This suggests that there may be a barrier to the passage of glucose from mother to fetus. But the gradient may be much less than it appears, for a significant proportion of the glucose is consumed during passage to meet the large energy requirements of the placenta.

The amount of glucose reaching the placenta is the product of the concentration in the blood and the volume of blood passing through the placenta. Deficiencies in either or both can occur and may lead to retardation of growth. In many animals, lower blood glucose levels resulting from inadequate nutrition are usually at fault, but in the better-fed human subject, disorders of blood supply are more common.

Fetal regulation of glucose utilization. One wonders whether the rate of fetal growth is constantly restrained by availability of glucose, or alternatively, whether there is normally more than enough at all times, so that growth is restrained by the mechanisms that control glucose utilization. We will try to answer this question in a succeeding paragraph. But first we must consider the ways in which the fetus differs from the adult in its glucose balance.

As we have already seen, the levels of glucose in the fetus mirror those in the mother. Thus there is neither the need nor the opportunity for the fetus to regulate its own blood glucose concentrations. The mechanisms for doing so do develop in the fetus, but are largely dormant until birth, when the supply of glucose from the placenta ends abruptly. Nevertheless, two important processes are active from an early stage. The first of these, which we will discuss in more detail later in the chapter, is concerned with the storage of glucose as glycogen or fat to provide for the metabolic needs of the newborn until feeding begins. The second is the control of the rate at which glucose is used by the growing tissues, and this is probably owing to the action of insulin secreted by the fetal pancreas. When insulin is administered to the fetuses of experimental animals such as rats and sheep, the rate of growth is considerably accelerated despite low fetal blood sugar levels (hypoglycaemia), and there is an increase in the size of glycogen-storing organs such as the liver and heart as well as increased deposition of fat. A similar phenomenon can be observed in the overweight newborn babies of diabetic mothers (Fig. 3-1). In this disorder, lack of maternal insulin secretion causes elevation of blood glucose levels. Fetal blood levels rise with those of the mother and stimulate the release of insulin from the fetal pancreas, which leads in turn to increased growth and fat storage. Thus we are led to the conclusion that the rate of glucose utilization rather than glucose availability is the more important factor regulating growth under conditions of adequate nutrition.

The blood of many mammalian fetuses (including hoofed animals but not including higher primates) contains another

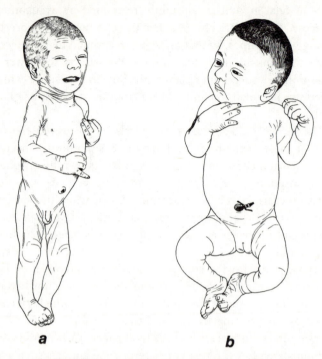

a   b

Fig. 3-1. Newborn babies who illustrate the effects of too little or too much insulin during fetal life. Baby *A* was diabetic at birth due to defective secretion of insulin by his pancreas. He shows signs of intra-uterine starvation because of limited ability to use glucose. On the other hand, Baby *B* is overweight and fat because of over-secretion of insulin by her pancreas leading to over use of glucose. Her mother was diabetic and blood glucose levels were elevated during pregnancy.

sugar, fructose, in concentrations often exceeding that of glucose. Fructose is synthesized in the placenta and is secreted into the blood stream of the fetus but not of the mother. The function which it subserves remains something of a mystery; little is metabolized by the fetus under usual conditions but more is used when glucose levels in the blood are low. Fructose may thus represent a placental mechanism designed to protect against the adverse effects on development of temporary shortages of glucose.

Cell growth

So far we have been concerned with growth of the whole body or of whole organs. Clearly, however, change in size of the body or of a particular organ is the result of increments in cell number and cell size. Cell multiplication is the most fundamental phenomenon of living matter, occurring to the greatest degree in fetal tissues, and when associated with the equally remarkable process of differentiation it forms the basis of development. Starting with a single cell, the fertilized egg, complete development of the human fetus to term involves 42 cell divisions, yet only a further five divisions are needed to attain adulthood.

The cells of the developing organism have a strong inherent capacity to multiply and to differentiate. Retarded growth due to a reduction in cell numbers is seen only in association with severe congenital anomalies or as a result of drugs such as aminopterin. Control of cell division and cell differentiation is exerted mainly within the tissues themselves by local 'organizers' synthesized within the cells in response to genetic information contained in their chromosomes. Endocrine secretions such as insulin and thyroxin are concerned more with the regulation of cell size than with cell numbers.

Fetal respiration

Respiration, by which is meant oxidation of foodstuffs (mainly glucose in the fetus) with the liberation of energy and carbon dioxide, is essential for the support of life and the processes of growth. Factors concerned with the supply of glucose to the cell have already been discussed and it is necessary now to turn our attention to the gases, oxygen and carbon dioxide.

During fetal life, the placenta, not the lung, is the organ of gaseous exchange and although anatomically these two organs appear so different, their respiratory functions have much in common. In essence, each consists of a membrane which is highly permeable to oxygen and carbon dioxide and through which the gases diffuse readily along a concentration gradient. In lungs, the gradient for oxygen is from the partial pressure of atmospheric oxygen (150 mm Hg) to that of oxygen in the blood. In the placenta, however, the gradient is from the partial pressure of oxygen in the maternal arterial blood (90 mm Hg) to that in fetal blood. The gradients are, of course, reversed for carbon dioxide. For various reasons the partial pressure of oxygen in the fetal blood cannot approach closely that in the maternal blood. For one thing, the placenta, which consumes considerable quantities of oxygen, lies between maternal and fetal circulations. Furthermore, the relationship of the fetal blood vessels to the maternal vessels is not a perfect counter-current system, and thus an equilibrium will be reached in the fetal blood which lies somewhere between the partial pressures of oxygen entering and leaving the maternal side of the placenta (Fig. 3-2). In fact, the partial pressure of oxygen (pO_2) in

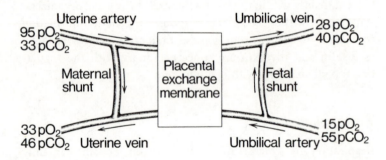

Fig. 3-2. Blood-gas exchange across the placenta. The figures are partial pressures of oxygen (pO_2) and carbon dioxide (pCO_2), expressed in millimetres of mercury.

fetal blood leaving the placenta is only about one third of maternal arterial pO_2. Nevertheless the universal success of pregnancy suggests that the fetus suffers no disadvantage from a low arterial pO_2. The fetus should not be thought of as chronically short of oxygen, nor, as some of the older physiologists did, equated with a man on the top of Mount Everest. The fetus is quite readily able to extract from its blood the volume of oxygen necessary for respiration, a volume in the human and sheep fetus near term of approximately 15 ml/min. This is possible mainly because of a high cardiac output ensuring a large blood flow both to the placenta and to the fetal tissues. In addition, late in gestation in some species the oxygen dissociation curve is shifted to the left of that of the mother, and this allows more oxygen to be carried at the same pO_2.

The fetus appears to lack control over the amount of maternal blood perfusing the placenta. Nevertheless the pregnancy hormones, oestrogen and progesterone, have ensured that the blood supply is adequate by promoting growth of the blood vessels throughout pregnancy. Towards term in man, at least 10 per cent of maternal cardiac output reaches the placenta; thus, in normal circumstances, the oxygen supply to the fetus is not only adequate to provide for the requirements of growth, but also has sufficient reserve to permit fetal survival when uterine blood flow is intermittently interrupted by the uterine contractions of labour. But in some circumstances, a deficient maternal blood supply may lead to retardation of growth or even to fetal death. In man, for example, the maternal blood vessels may be damaged by complications of pregnancy such as toxaemia.

Placental growth

A discussion of fetal growth would be incomplete without including the placenta, for not only is the placenta largely composed of fetal tissue, but also there is a strong correlation between the size of the fetus and the size of its placenta. Not enough is known about the factors determining placental growth to be sure whether the fetus has the means to influence the growth of the placenta, or whether placental size might limit fetal growth as a result of limitation of transfer of metabolic fuel and building materials (carbohydrate, amino acids and oxygen), or even whether growth of the placenta and the fetus simply proceeds hand in hand. Certainly, experimental reduction in placental mass, as George Alexander achieved in sheep when he surgically excised uterine caruncles prior to pregnancy, can cause reduction in fetal weight. But we should be unwise to infer from these observations that the normal placenta exercises a restraining influence on fetal growth.

FETAL FUNCTIONS

The development of function, whether it be in the muscles of a weight-lifter, the skilled fingers of a violinist or the ability to think and reason, depends on use; disuse leads to decay. It is inconceivable that the whole gamut of functions apparent in a newborn baby was not present moments earlier before birth, and was not in frequent use during much of intra-uterine life. A realistic mental picture of the dynamic state in which the fetus spends its intra-uterine existence can be obtained by imagining an experiment in which a vigorous newborn animal is re-attached to its placenta and replaced within the uterus. Of course the change in environment would require certain adaptations. In particular, air would no longer be available to breathe and the placenta must substitute for the lungs as the organ of gas exchange. An alternative supply of food, too, must be found, but the placenta is well capable of substituting for the mammary glands. There would be a sudden ease of movement though limited by the confines of the uterus, as the almost insupportable weight of body changed to weightlessness on total immersion in the amniotic fluid. The hard work of maintaining body temperature at 15 °C or more above the temperature of the air would disappear as the newborn re-entered its thermo-statically controlled waterbath. Our experimental animal would also have to make many subtler adaptations as it returned to fetal life. In the remaining paragraphs we will discuss, system by system, the special features of function that distinguish fetal from extra-uterine life.

Cardio-vascular functions ①

The circulatory system of the fetus is essentially that of the adult adapted to the special circumstance that the placenta functions for the lungs as the organ of respiration. This may sound like a relatively minor difference, but the adaptation is a feat that any hydraulics engineer would be justifiably proud to accomplish. It entails the diversion of almost the entire output of the right ventricle of the heart from the lungs into the umbilical circulation; and what is more, the diversion is achieved in a way that allows almost instantaneous conversion to the adult type of circulation as the first breath is taken. The anatomical solution to the problem is straightforward – the pulmonary artery and the aorta develop with a large connection between them so that blood heading towards the lungs is diverted directly into the aorta and thence to the placental circulation. The functional solution to the problem of redirecting blood to the lungs when the first breath is taken is also straightforward, though remarkably ingenious. The diversionary passage, the ductus arteriosus, has a muscular wall that is sensitive to the partial pressure of oxygen in the blood and responds to an increase in oxygen tension by contracting and closing. Thus, when arterial oxygen tensions rise with the onset of breathing, the ductus will tend to close. It is aided in this by an abrupt rise in the pressure in the aorta when the umbilical circulation, which has acted as a low resistance shunt between the arterial and venous systems, is removed (see Fig. 3-3). Before birth there is little blood going to the lungs and therefore little returning from the lungs to enter the left atrium of the heart. In the adult system this would mean that almost all of the venous return to the heart, including blood from the placenta, would enter the right atrium. In the fetus the difficulty is overcome by the presence of a hole in the septum between the right and left atria allowing half of the blood entering the right atrium to pass to the left atrium. The hole, the foramen ovale, is cunningly designed as a flap valve that not only

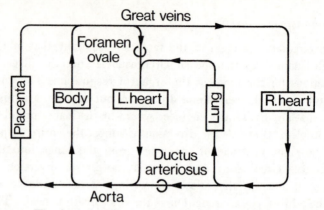

Fig. 3.3. The fetal circulation. This diagram differs from that for an adult in showing a placental circulation, a ductus arteriosus and a foramen ovale.

diverts the better-oxygenated blood of the inferior vena cava into the left atrium and thence to the brain, but also shuts at birth when the blood pressure in the left side of the heart rises to a level above that of the right side. Circulatory adaptation of the fetus thus depends on two diversionary passages, one a sphincter sensitive to oxygen tension, the other a flap valve sensitive to pressure, both of which close at birth when breathing begins.

Gastro-intestinal functions

Gastro-intestinal function in the human fetus has been extensively investigated by means of intra-amniotic injections, either of opaque media demonstrable by X-rays or of inert chemicals which can be transferred to the mother and excreted in her urine. In the latter part of pregnancy the fetus swallows about 500 ml amniotic fluid daily. When opaque media are included in the swallowed fluid it can be shown to pass rapidly into the stomach and onwards into the small bowel, reaching within a few hours the large bowel where it remains throughout the rest of pregnancy. Most of the swallowed material is water which is rapidly absorbed in the small bowel. Other materials of small molecular size such as electrolytes, glucose, urea and steroid hormones are also absorbed. Large molecules, cell debris, and sebaceous material secreted by the glands of the fetal skin are not absorbed, but accumulate in the large bowel, where, together with cells desquamated from the intestinal wall and bile pigments, they form the green faecal material known as meconium. Under normal circumstances defaecation *in utero* does not occur; but following acute oxygen deprivation, the large bowel may contract and the rectal sphincter relax, allowing the meconium to be discharged into the amniotic fluid.

We are not clear what benefit derives from the swallowing of such large amounts of fluid, but undoubtedly the practice helps to prepare the intestines for their digestive functions after birth. The swallowing of amniotic fluid may also contribute to fetal nutrition: approximately a gram of protein is swallowed daily and most of this is probably absorbed. Fetuses malformed by the failure of the oesophagus to develop as an open tube are of lower birthweight than normal babies, suggesting that the swallowed fluid has nutritive value. Probably the most important function of swallowing is concerned with control of amniotic fluid volume. Excessive amniotic fluid is an invariable accompaniment of disorders of fetal development that impair ability to swallow normally. This topic will be discussed further in the section on renal function.

The mouth of the newborn primate is much more than an orifice through which food is ingested; perhaps because of the ability to hold objects and to put them against the lips, its mouth has become a sensory organ of some importance in early postnatal life. It is not surprising, therefore, that some evidence of this aspect of the function of the mouth has been seen in the human fetus – X-rays not uncommonly reveal a fetus apparently sucking a thumb (Fig. 3-4)! The sensation of taste also seems to be present *in utero*. Experiments in which the rate of swallowing has been measured by determining the rate of disappearance from the amniotic fluid of red cells labelled with radioactive

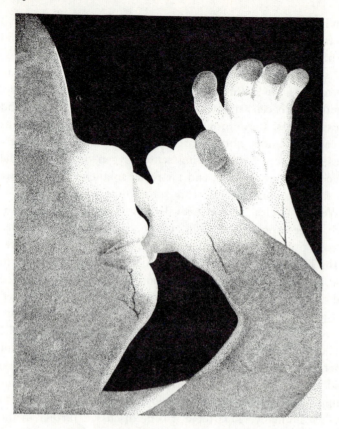

Fig. 3-4. A human fetus in a characteristically childlike posture; radiologists sometimes see this in X-rays of normal pregnant women. (Drawn from a photograph in E. Blechschmidt (1963) *The Human Embryo*. Stuttgart; Schattaeur.)

chromium have shown that the addition of saccharine to the amniotic fluid increases the rate of swallowing, whereas distasteful materials such as opaque media cause almost complete cessation of swallowing. The sucking reflex on which postnatal survival is so dependent is amongst the earliest of the coordinated reflex activities to appear in the developing fetus. In some fetal animals, strong sucking can be elicited before the middle of pregnancy by placing a fingertip between the lips.

Renal functions

Throughout pregnancy the placenta, not the kidney, is the organ of excretion. Waste products are transferred into the maternal circulation to be voided into the maternal urine. The fetal kidney might therefore seem to have no important function. Observations on fetuses, both animal and human, in which kidneys have failed to develop or in which there is complete obstruction of the urinary tract appear to bear this out; the fetuses survive to term and are born alive. However, such pregnancies are not entirely normal, suggesting that the fetal kidneys may have functions other than those concerned with excretion. Growth is retarded in fetuses with renal agenesis and also in fetal lambs from which the kidneys have been removed earlier in pregnancy. Paradoxically, growth hormone levels in the blood of the fetal lambs become abnormally elevated after removal of kidneys. One suggested function of the kidney is to convert the large protein molecule of growth hormone to a smaller, biologically active molecule.

Another anomaly observed in pregnancy when the fetus lacks kidneys is the virtual absence of amniotic fluid. The mature normal human fetus may excrete 450 ml of urine daily, a volume that would seem a significant contribution to a total amniotic fluid volume of about 1500 ml. But the hourly turnover of water in the amniotic fluid is 300–600 ml, so the contribution of urine is really very small. Fetal urine has been suggested as an important source of amniotic fluid protein, which provides the osmotic pressure necessary to retain water within the amniotic sac. Loss of protein from amniotic fluid, as we are already aware, occurs by fetal swallowing. The total mass of amniotic fluid could be regulated by the balance between fetal swallowing and fetal micturition. Aberrations of amniotic fluid volume associated either with defective fetal swallowing or defective micturition certainly are consistent with this (Fig. 3-5). However, the amount of protein in fetal urine is too little to account for anything but a small part of the total protein

The fetus and birth

turnover in amniotic fluid and although total protein mass is likely to be the major factor regulating amniotic fluid volume, neither the source of the protein nor the mechanisms controlling its entry into amniotic fluid are known.

In those mammals with a persistent allantoic sac (this does not include humans), the fetus can void urine either through the urachus to the allantoic sac or through the urethra to the amniotic sac. In early pregnancy most of the urine probably goes to the allantoic sac, but as pregnancy progresses more is voided into the amniotic sac.

Fetal breathing movements

Rhythmic movements of the chest, comparable to those of breathing, have usually been thought to be absent except in

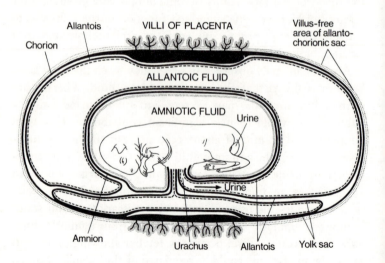

Fig. 3-5. Diagram of the membranes of a fetal kitten in an early stage of pregnancy, before obliteration of the yolk sac. Urine enters the allantoic sac through the urachus and the amniotic sac through the urethra. (From E. C. Amoroso, Placentation, in *Marshall's Physiology of Reproduction*, vol. 2. Ed. A. S. Parkes. London; Longmans (1952).)

fetuses *in extremis* as a result of deprivation of oxygen or accumulation of carbon dioxide. Considering the need to establish vigorous and effective respiratory movements immediately after birth it would be surprising if normal intra-uterine life presented no opportunity for the exercise of respiratory muscles. Recent observations have shown that rhythmic breathing movements do, in fact, occur at least during the last third of fetal life in the lamb. The movements are quite vigorous, causing negative pressures of up to 25 mm Hg within the chest, and having a frequency of 1–4 per second. They may be present for as much as 8 out of each 24 hours and may continue for more than 30 minutes at a time. What is even more remarkable and quite unexpected is the observation that the movements occur during rapid eye-movement (paradoxical or dreaming) 'sleep'. During periods of 'wakefulness' when there is active skeletal movement, or during 'deep sleep', breathing movements are absent.

In the course of X-ray investigations of the human fetus, when opaque media are injected into the amniotic fluid, the material is never detected in the fetal lungs. This has been interpreted as an indication that breathing movements are absent in the human fetus. However, despite their strength, fetal breathing movements in the lamb cause little net shift of fluid within the respiratory passages; a valve, probably at the level of the glottis, effectively prevents the entry of amniotic fluid into the trachea. What small flow of fluid there is consists of an out-flow of accumulated lung fluid at a rate of approximately 2 ml per hour. There is no reason to believe that human or other mammalian fetuses differ from the lamb in their breathing movements.

Endocrine functions

The endocrine organs are functional by the end of the first quarter of pregnancy in the human fetus and thereafter the fetus is self-sufficient in its hormonal requirements. In general,

The fetus and birth

the fetal glands secrete the same hormones with the same functions as those of the adult. But this information conveys nothing of the complexity and fascination of fetal endocrinology, for not only do the fetal organs subserve a number of unique functions vital to normal development, but there are also complicated inter-relationships between mother and fetus.

A detailed description of one of the most interesting of the fetal endocrine systems, the feto-placental unit, will be found in Book 3. This unit, found in man and probably some other mammals, is so named because the complete elaboration of certain hormones (particularly oestrogens) secreted by the placenta needs the combined efforts of both the placenta and of fetal organs. Enzymes concerned in the biosynthesis of the hormones are distributed amongst the various tissues so that no one organ is capable of complete synthesis. Of the many and varied special problems in fetal endocrinology apart from the feto-placental unit, it is only possible within the scope of this chapter to cite one or two examples relating to each of the endocrine organs. But first, let us state some principles that describe general hormonal relationships between the mother and the fetus.

1. The fetal endocrine system profoundly influences the mother. The hormones secreted by the feto-placental unit into the maternal circulation are responsible for most of the physiological changes that occur in the mother during pregnancy, changes involving every system of her body. For example, alterations in the function of the maternal central nervous system are apparent in altered behavioural patterns; her heart output is increased and her peripheral blood vessels dilated; in the kidneys there is an increase in glomerular filtration rate; the blood levels and secretion rates of several maternal hormones are increased; the oestrous cycle is suppressed; glucose, fat and protein metabolism is altered; ligaments relax; and the mammary glands develop. In mammals that are not dependent on the corpus luteum for maintenance of pregnancy (including man, horse, sheep and guinea pig) the changes are induced directly by placental hormones, while in those species that require a functional corpus luteum throughout the whole of pregnancy (including rabbit, pig, rat and goat) the influence of the conceptus is exerted indirectly through hormones that regulate corpus luteum function.

2. *The maternal endocrine system does not directly influence the fetus except in disease.* The protein and polypeptide hormones of the pituitary and pancreas are completely excluded from the fetus by their inability to pass through the placenta. Hormones of smaller molecular size such as the adrenal corticosteroids can pass through the placenta of some species (e.g. man) but not others (e.g. sheep); but even when passage does occur much of the active hormone is metabolized to biologically inactive material by specific enzymes in the placenta. In those species that have been investigated, thyroxin has been found to pass through the placenta slowly but in insufficient amounts to provide for the needs of mature fetuses. Alternatively, a maternal endocrine organ may influence its counterpart in the fetus by indirect means; insulin is a good example of this (see below).

3. *The fetal endocrine system has certain specialized functions not present in adult life.* At certain critical phases of development fetal hormones have special functions never again required of them. Examples such as sex determination, preparation for birth and initiation of labour are described in later sections of this chapter.

The hypothalamus. The maternal hypothalamus is described in detail in Book 3. Here we should remember that the hypothalamus is active during fetal life and that it co-ordinates the activities in the other endocrine organs in the same way that it does in the adult. In one way, however, it is strikingly different from the adult hypothalamus – in the course of maturation the function of the fetal hypothalamus can be permanently

The fetus and birth

modified by brief exposure to certain hormonal stimuli. This is discussed more fully in the preceding chapter. Other intra-uterine factors may possibly leave imprints on the hypothalamus that could permanently modify behaviour. This is one good reason why drugs, even those that have no teratogenic effects, should be used with caution in pregnant women.

The pituitary gland. The anterior pituitary gland is functional throughout most of fetal life and it secretes the same trophic hormones that are secreted by the adult gland. In general, the trophic hormones have functions similar to those of the adult, stimulating both secretion and growth in the target organs. Fetal hypophysectomy prevents normal development of the thyroid, adrenals and testes, but the pancreas and ovaries are little affected.

When the fetal pituitary is separated from the hypothalamus by dividing the pituitary stalk, the signs of defective pituitary function are much less distinct than after hypophysectomy, suggesting that the fetal pituitary is less dependent than the adult pituitary on releasing factors from the hypothalamus. Nevertheless, the usual feedback mechanisms act on the hypothalamus to regulate trophic hormone release from the pituitary. Administration to the fetus of cortisol, the major secretory product of the adrenal cortex, causes adrenal atrophy by inhibiting the release of corticotrophin (ACTH). This effect of cortisol is shown particularly clearly in human pregnancy when corticosteroids are administered to the mother. Passage of the corticosteroid across the placenta causes a sharp reduction in adrenal secretion of the precursor material used by the placenta in the biosynthesis of oestrogens. Consequently there is a steep fall in the excretion of oestrogens in the maternal urine.

The thyroid gland. Thyroxin is particularly important in the normal development of the brain, the bones and the hair or wool. Newborn animals suffering from the effects of hypo-thyroidism *in utero* may be of normal size, but show behavioural retardation, have a bone age that has lagged far behind the chronological age, and are deficient in body hair. These changes are present, even though maternal levels of thyroxin are normal, because the restricted passage of thyroxin across the placenta allows insufficient entry of maternal thyroxin to meet fetal needs. A very complicated feto-maternal endocrine relationship may develop in the course of the human thyroid disease, thyrotoxicosis. The manifestations of this disorder are the result of excess secretion of thyroxin from a gland that is under constant stimulation by an abnormal antibody called long-acting thyroid stimulator (LATS). Although thyroxin crosses the placenta with difficulty, LATS crosses more readily, causing abnormal activity in the fetal thyroid and hence fetal thyrotoxicosis. Furthermore, the thyroid-blocking drugs used in the treatment of maternal thyrotoxicosis also cross the placenta and if used injudiciously may suppress function in the fetal gland. Not only may the fetus then show evidence of hypo-thyroidism, but also it may develop a goitre because low thyroxin levels lead to hypersecretion of thyrotrophin from the fetal pituitary.

The fetal pancreas. Of the various hormones concerned in the regulation of growth, insulin from the fetal pancreas is probably the most significant because it regulates the rate of glucose utilization which, as we have already seen, is the source of energy for fetal growth. It is intriguing to wonder how the release of insulin from the fetal pancreas is regulated. In the normal adult the insulin response to taking food is deter-mined in two ways: first, a hormone that stimulates insulin secretion is released from the lining of the intestine when glucose in the ingested food comes into contact with it. Consequently, a rise in the concentration of insulin in the blood stream precedes the rise in glucose. Second, glucose has a direct stimulatory effect on the pancreas. In the fetus, insufficient glucose is swallowed to activate the intestinal hormone system. Further-more, the fetal pancreas responds to elevations in blood glucose levels with only a small rise in insulin secretion. Thus, insulin secretion in the fetus seems likely to be continuous and relatively unvarying. Nevertheless, under abnormal circumstances the fetal insulin secretion can be abnormally high. In diabetic women who have elevated blood glucose levels over long periods of time, excessive growth of the fetal pancreas occurs, no doubt in response to a sustained elevation of fetal blood glucose levels (Fig. 3-6). The consequences of increased insulin secretion from the fetal pancreas on the rate of growth have already been discussed.

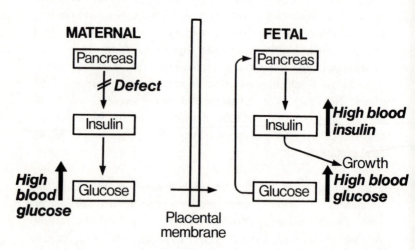

Fig. 3-6. Feto-maternal interrelationships in diabetes. Deficiency of insulin in the mother leads to elevated glucose levels in her blood which leads in turn to elevated glucose levels in the fetal blood stream, increased secretion of insulin from the fetal pancreas and accelerated growth.

The fetal gonads. The fetal ovary in most mammals is small and shows little evidence of activity. In the preceding chapter on sex determination the passive role played by the ovary has been described. The extraordinary phenomenon of ovulation that occurs in the fetal giraffe is likely to be of an incidental nature, serving no useful purpose. The fetal testis, on the other hand, is a highly active endocrine organ whose function is essential for normal sexual differentiation in the male. The development and function of the testis is regulated by the hypothalamus and pituitary, as can be readily demonstrated by observing the effects of fetal hypophysectomy early in pregnancy. The interstitial cells of the testis are also stimulated by human chorionic gonadotrophin (HCG) and it is tempting to link the peak of HCG secretion at 60–70 days with the time of sexual differentiation in the human fetus.

The fetal adrenal glands. No particularly important function has been attributed to the medulla of the fetal adrenal. Like the adult adrenal medulla it secretes both adrenalin and noradrenalin. It differs from the adult gland in containing a higher concentration of noradrenalin than adrenalin. In addition there is a direct response to asphyxia as well as the adult-type response mediated by the splanchnic nerves. The release of catecholamines from the adrenal medulla probably aids survival during birth asphyxia by contributing to the maintenance of the circulation and by mobilizing glucose from body stores.

The cortex of the adrenal is the most versatile of the fetal endocrine glands. The large number of specialized functions attributed to it fall into two groups. The first group includes actions that have in common timing of developmental events; all of these depend upon the ability of cortisol to induce activity in specific enzyme systems. Such cortisol-inducible enzymes have been identified in a variety of fetal organs including brain, liver, eye, intestine, adrenal medulla, placenta and lung. The developmental consequences of 'switching-on' an enzyme system may be profound and some of the physiological events that follow are described below in the sections entitled 'Preparations for birth' and 'Control of parturition'.

The second group of functions of the adrenal cortex depends on the development in certain species of a specialized area of the cortex known as the fetal zone. This reaches its most highly developed form in the human fetus where its bulk is so great

The fetus and birth

that the total weight of fetal adrenal tissue is equal to that of the adrenal from a child of 10–12 years. After birth the fetal zone regresses quite rapidly and by 2–3 months of age has disappeared. Biochemically, the zone is distinguished by its inability to synthesize cortisol. Instead, large quantities of dehydroepiandrosterone sulphate, the precursor of oestrogen, are secreted into the fetal circulation. Further details of this aspect of adrenal function can be found in Book 3, Chapter 1, in the section describing the feto-placental unit.

Immunological functions

There are many remarkable mechanisms involved in successful reproduction but none more so than the means by which the conceptus is able to avoid immune rejection by the mother. The mammalian fetus is attached to the uterine wall in a way that resembles a graft, and yet genetic differences exist between the mother and her fetus. Half the genetic makeup of the latter is derived from the mother, and half from the father. The maternal component of the fetus poses no problem of incompatibility, but the paternal component is foreign to the mother and might be expected to lead to rejection. But even transplanted eggs, in which both the maternal and paternal components are foreign to the donor mother, can implant and develop normally. How this tolerance is achieved is an unresolved mystery. (These and other aspects of the immunology of the fetus are discussed in the first chapter of this book and in Book 4, Chapter 4.)

Neurological functions

In the young animal we are used to the idea that the development of brain function for motor activities, for perception and for behavioural patterns is to a great extent dependent upon, and modified by, environmental influences; genetic pre-programming seems less in evidence. That is not to say, of course, that such pre-programming is absent. To take a simple

94

example, a group of young animals reared in isolation from their elders will develop the general patterns of social behaviour characteristic of the species. In the fetus, on the contrary, the development of brain function follows a course that is mainly genetically determined and is largely independent of environment and of peripheral events within the fetus. In mammals with long gestation lengths the sensory organs of the fetus are highly developed and are capable of response to the usual stimuli. Although the intra-uterine environment is by no means the sensory void pictured by physiologists of old, neither is the scope or intensity of stimuli likely to play an important part in brain development. The intensity of sound, for instance, within the human uterus is approximately 50 decibels (comparable to sound intensity in a quiet office) and the intensity of light (unless the mother happens to be sunbathing in a bikini) is similar to that of a darkened lecture room. Thermal stimuli are almost completely absent, the intra-uterine temperature being maintained within very narrow limits at about 0.5°C above that of the mother. Tactile stimuli are minimized by the cushioning effects of the amniotic fluid, and the state of near-weightlessness favours a low input of proprioceptive information. Nevertheless, the fetus can readily be shown to be responsive to unaccustomed levels of any of these stimuli. A loud noise close to the uterus, a flash of high intensity light, the touch of a needle tip on the skin, a rapid change in the position of the mother, or the injection of a cold fluid into the amniotic sac, all may evoke a quick and vigorous reaction in the form of fetal movement. At the present time there is little evidence that would support those psychologists who would have us believe that maternal behaviour can have effects on brain development that play a significant part in moulding the personality of the child and adult. Imprinting can and does occur *in utero* (e.g. effects of androgen on the hypothalamus) but is the result of genetically controlled events within the fetus rather than of random events in the environment.

To the mother, whether human or animal, fetal movements are her only certain evidence of her pregnant state. They start

95

almost as soon as the limbs are formed, but until the fetus becomes bigger, movements do not make their presence felt. That they can be felt by the human mother as early as the 14th week is an indication of their vigour. The purpose served by fetal movements is not altogether clear. Part of their function is concerned with the development of muscle action by 'exercise' but fetal muscles develop, at least anatomically, in the absence of nervous connections. Indeed, in species such as the rat, the muscles are well formed before nerves have reached them, and motor end-plates begin to develop only about four days before birth. In mammals that have longer gestation periods and that are born in a more mature state than the rat, motor end-plates appear during the first half of pregnancy, allowing a greater opportunity for the motor nerves to dominate further maturation of the muscles.

If a fetal lamb, still attached to the mother by the umbilical cord, is observed in a warm water bath, three states of movement can be distinguished. In the first state, called 'deep sleep', there is little or no spontaneous movement and the response to stimuli is sluggish. In the second state, termed 'wakefulness', the lamb opens its eyes, seems aware of its surroundings, responds briskly to stimuli and there is much spontaneous movement of the limbs and trunk. The movements probably have little to do with what is happening in the brain and are probably the result of spinal reflex activity. Indeed, headless monsters are remarkable for their abnormally high motor activity, suggesting that the brain may have inhibitory rather than excitatory effects. The third state, 'rapid eye-movement sleep' (REM sleep) is characterized by twitching movements of the legs, trunk and facial muscles and by rapid horizontal movements of the eyes. Stimuli usually have no effect other than to cause cessation of the twitching movements.

REM sleep is an intriguing phenomenon that has so far defied a satisfactory explanation. In adult individuals deprived of sleep, restoration of the deficit of REM sleep gets priority over other sleep states, suggesting that it has functions of some importance. One hypothesis suggests that during REM sleep the brain 'practises' its more complex perceptive and motor functions. This idea has something to commend it in the fetal lamb which spends at least a third of the day in this form of sleep; three of the most highly co-ordinated motor activities that occur after birth – eye movements, breathing movements and movements of the mouth – are present more or less continuously through REM sleep.

The development of function in the fetal brain is the darkest corner in fetal physiology. New tools such as fetal electroencephalography are now beginning to allow study of the undisturbed fetus *in utero* and can be expected to advance our knowledge rapidly.

PREPARATIONS FOR BIRTH

At birth, separation from the placenta suddenly deprives the newborn of the supplies of oxygen and glucose to which it has been accustomed. Unless an alternative supply is found rapidly, death will ensue. In normal circumstances breathing starts within seconds of birth, and is fully established after a few minutes. Glucose is rapidly released from carbohydrate reserves, particularly from glycogen stored in the liver and heart muscle. This source of glucose is exhausted within a few hours and fat then becomes the main metabolic fuel, supplying both fatty acids and glycerol for oxidation in the tissues. Even the fat stores can sustain life for only a limited period, but this is usually more than sufficient to bridge the variable gap until suckling starts. Successful transition from intra-uterine to extra-uterine existence is thus clearly dependent on events that have preceded birth – functional maturation of the lungs, laying down of carbohydrate and fat reserves, and the onset of lactation.

Maturation of lung function

Given reasonable anatomical development, the lungs can be

97

maintained in an air-distended state after birth provided that an adequate concentration of a surface active material called 'surfactant' is present in the lung alveoli. This material is secreted by the membrane lining the alveoli and its first appearance more or less coincides with the period of viability, i.e. the earliest time at which extra-uterine survival is possible. A surface film containing surfactant has the unique property of exerting a high surface tension when it is stretched, but a very low tension

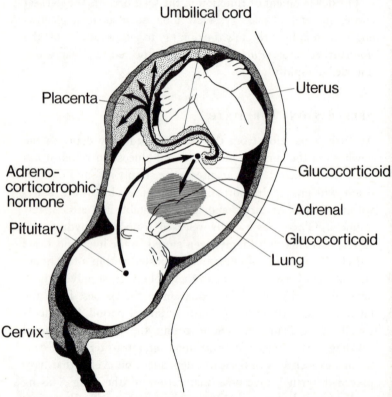

Fig. 3-7. The fetal adrenal cortex occupies a key position in regulating many of the physiological processes that precede birth. Cortical hormones (glucocorticoids) influence the time when labour starts and also prepare the lungs for breathing by stimulating the formation of a surface-active compound. (From G. S. Dawes, Hazards of birth, *Sci. J.* **6**, 86 (1970).)

when the surface is compressed. As a result the surface tension in a small alveolus is less than that of a larger alveolus and their volumes will tend to equalize, thus preventing lung collapse and promoting alveolar stability. The time of first appearance of surfactant may be related to activity of the fetal adrenal. Experiments in fetal lambs and fetal rabbits have shown that stimulation of the adrenals by ACTH injections at a time when surfactant is absent from the lungs results in rapid appearance of surfactant in high concentration. Fetuses delivered prematurely after such treatment show a remarkable ability to breathe and to survive (Fig. 3-7). A period of intense adrenal activity of the fetal adrenals precedes birth in a wide variety of mammals and we may reasonably suppose that one of the purposes of this increased corticosteroid secretion is preparation of the lungs for birth.

Carbohydrate and fat reserves

During the latter part of gestation the fetus of many species accumulates large amounts of glycogen, particularly in the liver. As long as adequate supplies of glucose can reach the fetus from the mother, the total amount of carbohydrate stored in the various tissues appears to depend mainly on the level of fetal adrenal activity. In several species experimentally induced hypofunction of the adrenals is associated with a big reduction in liver glycogen content which can be restored by administering a corticosteroid. Moreover, when ACTH or corticosteroids are injected into normal fetuses, liver glycogen content becomes abnormally high. The enhanced rate of storage that normally occurs preparatory to birth is probably related in part to accentuated adrenal activity near term.

The pronounced tolerance of newborn animals to severe oxygen deficiency (hypoxia) is attributable, amongst other things, to the high concentration of glycogen present in cardiac muscle. This glycogen store allows the heart to continue its contractile activity in the absence of oxygen, by means of anaerobic gly-

The fetus and birth

colysis. The brain is in a less fortunate position, having no glycogen reserves and having to rely on circulating blood glucose. Neonatal hypoglycaemia is a not uncommon condition in human beings and is usually seen in babies who show signs of intra-uterine growth retardation. These babies are born with abnormally low levels of liver glycogen, presumably because they have been deprived of glucose by poorly functioning

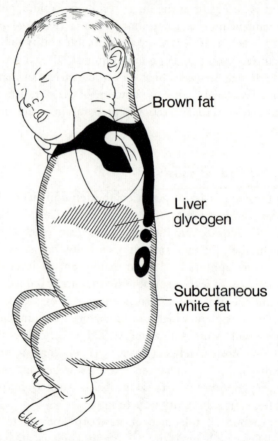

Fig. 3-8. Major sites of energy-stores in the fetus. Glycogen provides for the immediate needs of the newborn baby but may be exhausted rapidly. White fat is then metabolized until feeding is established. Brown fat acts as a source of heat thus sparing glycogen.

placentas. The brain may be irreparably damaged during episodes of hypoglycaemia, especially if they are accompanied by hypoxia.

Storage of fat is regulated by insulin rather than by corticosteroids. When nutrition is good and blood glucose is maintained at relatively high levels the action of insulin ensures that glucose available after the requirements of growth are fully met is diverted into fat stores. But when glucose supplies are restricted, the needs of growth have priority over storage and the newborn animal has an emaciated appearance.

As well as the usual white fat, most fetuses have substantial deposits of a peculiar form of adipose tissue called brown fat (Fig. 3-8). Whereas white fat liberates fatty acids and glycerol into the circulation to be metabolized in tissue elsewhere, brown fat is oxidized and releases its energy in the form of heat largely within itself. It plays an important part in maintaining body temperature after birth when the newly born animal is no longer within the neutral thermal environment of the mother.

Initiation of lactation

In most mammals, man being an exception, lactation precedes labour by a variable number of days, thereby enabling the young animal to obtain substantial quantities of food soon after birth. The mechanisms controlling the onset and maintenance of lactation are fully described in Book 3. But it should be emphasized at this point that the hormonal changes responsible for the onset of lactation emanate from the fetus and placenta and thus can be regarded as a further example of preparation by the fetus for its birth. The fetal mechanisms controlling the initiation both of lactation and of parturition probably have much in common.

CONTROL OF PARTURITION

According to the usually accepted view of the hormonal control of pregnancy, the contractile activity of the uterus that would otherwise expel the growing conceptus is inhibited by progesterone. Then, at term, the inhibitory action of progesterone is withdrawn, allowing uterine contractions to develop to the point where the cervix of the uterus dilates and labour progresses to delivery. In early pregnancy the source of progesterone is always the corpus luteum of the ovary, but later in pregnancy, as we have already discussed, this function of the corpus luteum may be taken over by the placenta of some mammals. Thus mammals may be divided into two groups; in one, the maintenance of pregnancy is dependent upon corpus luteum function throughout the whole of pregnancy, whilst in the other, dependence on the corpus luteum ceases at some point during pregnancy. In general, species having short gestation lengths belong to the former group, and those with long gestation lengths belong to the latter group, but there are exceptions. Viewed in this way, fundamentally different mechanisms controlling parturition could be expected to operate in the two groups, the mechanism of one group being concerned with control of corpus luteum function, the other with placental function. Attitudes of this sort have dominated thinking for more than a generation since progesterone was first discovered. We now prefer to believe that a common mechanism applies to most, if not all, mammals and that species differences express themselves in variations on the theme rather than in distinctly different compositions. Underlying this altered outlook is the appreciation that it is the conceptus (or the fetus itself), and not the mother, that exercises the dominant control of parturition.

Control of parturition in placenta-dependent mammals

Much of the new information about the control of parturition has come from studies in the sheep, an obliging animal not only bearing one or two fetuses of a large enough size to make intricate intra-uterine surgical procedures technically feasible, but also having a gestation length (147 days) sufficiently long to allow extended chronic experiments. Out of this work has emerged a picture which, although incomplete, is reasonably clear and can serve us as an example to be tested in other mammals.

The clues that led to the discovery of the well-kept secret that the fetus, not the mother, controls labour in sheep, came from sheep farmers in the mountains of Idaho. They noticed that sheep grazing in mountain pastures at certain times of the year often had greatly prolonged pregnancies ending with death during attempts to deliver lambs weighing two or three times as much as normal. The cause of the disorder was eventually traced to a weed, *Veratrum californicum*, on which the ewes were feeding in early pregnancy. An alkaloid in the weed, although harmless to the mother, could severely damage the developing fetal brain. These observations suggested that the normal fetal lamb participates in the mechanisms controlling labour to an extent that labour could fail to occur when the fetal contribution was missing. Now, in the postmature lambs of the Idaho sheep, the fetal pituitary was invariably missing or displaced and the adrenal glands were underdeveloped, and these things pointed the way for investigators to elucidate the fetal mechanisms in the laboratory. Indeed, experiments soon confirmed the suspicions that the abnormality of the fetal pituitary and adrenals was causally related to the disordered control of gestation length. Surgical removal of the pituitary or of the adrenals from normal fetal lambs resulted in pregnancies that continued for weeks beyond term without any signs of labour. The same effect was obtained when the stalk connecting the hypothalamus with the pituitary was divided, suggesting that the hypothalamus as well as the pituitary and adrenals was implicated.

Further experiments showed that the chain of events leading to parturition at term begins in the fetal hypothalamus. The next link is the fetal pituitary which responds to hypothalamic activity by releasing ACTH. The fetal adrenal, in turn, is stimulated to secrete more corticosteroids. The fetal mechanism proved to be such a potent one that premature labour could be induced at any time in the second half of pregnancy by artificially stimulating the pathway with fetal infusions of either ACTH or a corticosteroid.

As we saw earlier, the sheep belongs to the group of mammals that synthesizes progesterone in the placenta and that maintains pregnancy after removal of the ovaries in the second half of the pregnancy. But the sheep differs from some other members of the group in showing a sharp fall in the concentration of progesterone in the maternal blood immediately before the onset of labour. Accordingly it has been felt that withdrawal of the inhibitory effects of progesterone on the uterine musculature, the myometrium, was the major factor controlling labour. This seemed even more likely when the same fall in progesterone was found to precede premature delivery induced by stimulation of the pituitary–adrenal system of the fetus. These observations were placed in a different perspective when neither labour at term in normal ewes nor premature labour in ewes carrying fetuses infused with ACTH, was prevented by administering progesterone to the ewes in doses that compensated for any fall in progesterone concentration. This raised the possibility that the withdrawal of progesterone played a role of secondary importance compared with another factor, as yet unknown. Recent investigations strongly suggest that the latter factor is a prostaglandin.

The prostaglandins are a group of long-chain fatty acids, first identified more than thirty years ago, but shown only in the last few years to have complex actions in every organ of the body. Two of them, $PGF_2\alpha$ and PGE_2, are powerful in initiating parturition. Minute quantities of either lipid cause strong uterine contractions in a variety of experimental animals and in man. In fact, the uterine response is so sensitive that it has been used for many years as the basis of a bioassay.

A search for $PGF_2\alpha$ and PGE_2 in tissues of the fetus and the ewe during experiments in which premature delivery was induced by fetal stimulation was rewarded by finding that a great increase in concentration of $PGF_2\alpha$ occurred in the maternal cotyledons, in venous blood draining from the uterus,

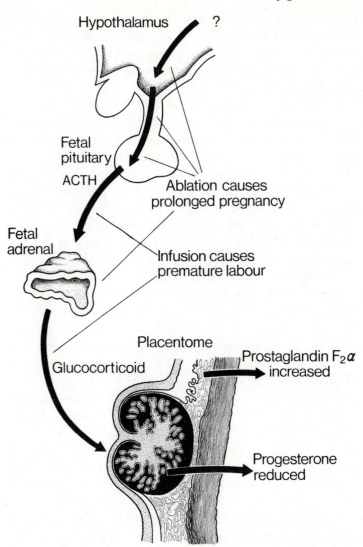

Fig. 3-9. Diagrammatic summary of a hypothesis explaining how the fetal lamb controls the onset of labour. Experimental procedures that lengthen or shorten pregnancy are shown. (From G. C. Liggins, The foetal role in the initiation of parturition in the ewe. In *Foetal Autonomy*. Ed. G. E. W. Wolstenholme and M. O'Connor. London; Churchill (1969).)

and in the myometrium. The rise in concentration in these tissues preceded labour by at least 24 hours and by the time labour started a five to tenfold increase had occurred. Little $PGF_2\alpha$ was present in the fetal cotyledon, either before or during labour.

Fig. 3-9 illustrates present views on the major factors controlling the onset of labour in sheep. The site of synthesis of $PGF_2\alpha$ is not yet known but, as suggested in the diagram, the site is likely to be the maternal cotyledon (decidua) from whence $PGF_2\alpha$ passes to the myometrium by an unknown route. Also to be elucidated is the means by which fetal corticosteroid controls prostaglandin synthesis in maternal tissues; if the action were a direct one, administration of big doses of ACTH or corticosteroids to the ewe should cause premature labour. This is not the case, except in the last few days of pregnancy.

As yet, there has been little experimental work in other mammals of the 'ovary-independent' group, to explore the extent to which fetal control of labour applies across species, although, however, fetal malformations involving the pituitary and adrenals are known to cause prolongation of pregnancy in both cows and women. The role of prostaglandins in the control of labour in other species is an even more unexplored field, but there are reasons for believing that prostaglandin may have an important place in the physiology of labour in women. $PGF_2\alpha$ and PGE_2 have been found in maternal decidua, amniotic fluid and blood during labour, but are absent before labour. As pharmacological agents, prostaglandins have proved more successful than the posterior pituitary hormone, oxytocin, in inducing labour or abòrtion at all stages of pregnancy.

Oxytocin is released from the maternal pituitary gland in the late stages of labour when more forceful uterine contractions are needed to expel the fetus. No increased release occurs at the onset of contractions; thus it is unlikely that maternal oxytocin is responsible for initiating labour. Human fetal blood at birth contains a high concentration of oxytocin and were it not for the fact that the placenta is rich in an enzyme, oxytocinase, that destroys oxytocin, we would have to consider the possibility that fetal oxytocin can cross the placenta to reach the uterine muscle in quantities sufficient to start labour.

Control of parturition in corpus-luteum-dependent mammals

Labour in corpus-luteum-dependent species seems clearly preceded by regression of the corpora lutea and a fall in the concentration of progesterone in the blood, and factors controlling luteal regression are evidently responsible for determining the time of onset of labour, but the nature of these factors is obscure. Maternal pituitary gonadotrophins are usually burdened with the responsibility for regulating corpus luteum function, and this view receives support from experiments showing that maternal hypophysectomy causes abortion. However, there are exceptions to this, and luteal regression follows disruption of the placentae as well as hypophysectomy. Maintenance of the corpus luteum is probably shared between the maternal pituitary and the conceptus; in the rat, the only species thoroughly investigated from this point of view, the conceptus appears to be more important than the maternal pituitary.

The goat, although a close relative of the sheep, is a corpus-luteum-dependent mammal; abortion occurs after maternal hypophysectomy or ovariectomy at any stage of pregnancy. It has therefore been of some interest to see whether the fetal goat exercises the same degree of control over parturition as the fetal lamb. The two species do have much in common: labour occurs in both when ACTH is administered to the fetus, and the same marked fall in the blood concentration of progesterone precedes premature labour in the goat as well as sheep. In the goat it is the ovarian, not the placental, secretion of progesterone that is affected by fetal adrenal activity, which means that a messenger of some sort must pass from the fetus to the ovary. In normal oestrous cycles the luteolytic factor released by the endometrium is now thought to be $PGF_2\alpha$. We may speculate that the same messenger may be released from the endometrium in pregnancy, leading to luteolysis and parturition.

If this speculation turns out to be correct, the common denominator in the mechanisms controlling labour in all mammals could be $PGF_2\alpha$; in corpus-luteum-dependent species, it would function mainly as a luteolytic agent, perhaps with an added oxytocic effect, whereas in other mammals it would have mainly an oxytocic action. The measure of control exercised by the conceptus over the synthesis and release of $PGF_2\alpha$ from the endometrium might range from little or none in marsupials, through control by both placenta and fetus in corpus-luteum-dependent species to fetal control in mammals with longer gestation lengths.

Circadian rhythms of birth

Not only the day of birth but also the time of birth has a physiological controlling mechanism in some mammals. For example, eight or nine out of every ten foals are born between 7.00 p.m. and 7.00 a.m. Similar, though much less pronounced, circadian rhythms in the time of birth can be recognized in women, pigs, mice and Chinese hamsters. Sheep, golden hamsters and rats on the other hand favour daylight hours. The purpose of the rhythms is lost in evolutionary history, but no doubt favoured survival of the newborn animal, perhaps by offering protection from predators. Whether the rhythm arises in the fetus or the mother is uncertain; since the rhythm depends on distinguishing light from darkness it is more likely to be determined by the mother. Neural mechanisms allow many mammals to suppress labour for limited periods of time when circumstances are unfavourable and the same pathways may mediate the circadian rhythm. One could say that the fetus may determine the day of birth, but the mother may determine its time.

In several ways our ideas about prenatal existence and the meaning of birth have changed radically in recent years. The notion of a dormant fetus springing to life upon entry into the world clearly needs as much revision as Wordsworth's thought that 'Our birth is but a sleep and a forgetting'. Development is remarkably complete before birth, both in form and function, and preparedness is the keynote.

SUGGESTED FURTHER READING

Some aspects of carbohydrate metabolism in pregnancy with special reference to the energy metabolism and hormonal status of the infant of the diabetic woman and the diabetogenic effect of pregnancy. J. D. Baird. *Journal of Endocrinology* **44,** 139 (1969).

Human fetal electro-cardiographic response to intrauterine acoustic signals. T. P. Barden, P. Peltzman and J. T. Graham. *American Journal Obstetrics and Gynecology* **100,** 1128 (1968).

Cellular growth, nutrition and development. D. B. Cheek, J. E. Graystone and M. S. Read. *Pediatrics* **45,** 315 (1970).

The alveolar lining membrane. J. A. Clements. In *Development of the Lung*. Ed. A. V. S. Reuck and R. Porter. Ciba Foundation Symposium. London; Churchill (1967).

The foetus as an allograft; the role of maternal unresponsiveness to paternally derived foetal antigens. G. A. Currie. In *Foetal Autonomy*. Ed. G. E. W. Wolstenholme and M. O'Connor. Ciba Foundation Symposium. London; Churchill (1969).

Foetal and Neonatal Physiology. G. S. Dawes. Chicago; Year Book Medical Publishers Inc. (1968).

Prenatal Environment. A. J. Ferreira. Springfield, Ill.; Thomas (1969).

Hormonal control of fetal development and metabolism. A. Jost and L. Picon. *Advances in Metabolic Disorders* **4,** 123 (1970).

The foetal role in the initiation of parturition in the ewe. G. C. Liggins. In *Foetal Autonomy*. Ed. G. E. W. Wolstenholme and M. O'Connor. Ciba Foundation Symposium. London; Churchill (1969).

Immunity in the foetus and the new-born. J. F. A. P. Miller. *British Medical Bulletin* **22,** 21 (1966).

Human fetal nutrition and growth. E. W. Page. *American Journal of Obstetrics and Gynecology* **104,** 378 (1969).

Three topics in placental transport. Amino acid transport: oxygen transfer: placental function during labour. M. Young. In *Foetus and Placenta*. Ed. A. Klopper and E. Diczfalusy. Oxford; Blackwell (1969.)

4 Manipulation of development
R. L. Gardner

Until recently, there had been few attempts to manipulate mammalian development, because it is more difficult than experimenting with developing frogs or sea urchins. Satisfactory methods had first to be devised for the culture of pre-implantation embryos and their successful transfer to uterine foster-mothers. Most of the experimental approaches have been adapted from the basic repertoire of embryologists interested in non-mammalian forms since the late nineteenth century. Their application to the mammalian embryo has proved an inviting challenge, and endeavours are now beginning to reap very considerable rewards. The aim of this chapter is to explain how some of these procedures may both enhance our understanding of mammalian development and confer a measure of control over it.

Experimental isolation of blastomeres and the fusion of cleaving embryos have already been discussed in detail in relation to problems of development in the early embryo (see Chapter 1). The mechanism underlying the differentiation of trophoblast and inner cell mass remains obscure, but the embryo is clearly divisible into these two groups of cells when it begins to secrete blastocoelic fluid. What are the properties of these tissues and how do they differ? Does the inner cell mass play any part in implantation? In what ways do the tissues interact in subsequent development? One must investigate the blastocyst by microsurgery in order to answer these and related questions.

The mouse blastocyst was chosen for this study. The availability of genetic and chromosome markers in this species outweighs problems posed by the small size of the blastocyst. The principal findings are described below; to what extent they may be generalized is conjectural at present, and broad inferences must await comparable studies in other species.

MANIPULATION OF THE BLASTOCYST

Comparison of isolated trophoblast and inner cell mass tissue

The $3\frac{1}{2}$-day blastocyst is roughly 0.1 mm in diameter and has therefore to be dissected under high magnification with the aid of a micro-manipulator. Blastocysts may be immobilized by suction on the flame-polished tip of a glass micro-pipette and then sectioned close to the inner cell mass with a micro-scalpel. This provides a source of trophoblast free of any enclosed cells of the inner cell mass (Fig. 4-1). Inner cell mass tissue is obtained most readily by tearing intact blastocysts open with pointed glass needles, and stretching the trophoblast out as a sheet. The inner cell mass is then gently scraped off the trophoblast (Fig. 4-1).

Trophoblast and inner cell mass tissue isolated in this way differ strikingly in their properties both in culture and following transfer to the uteri of pseudopregnant mice. These differences are summarized in Table 4-1. Formation of blastocoele fluid and induction of implantation are properties peculiar to the trophoblast, independently of inner cell mass tissue. However, the inner cell mass plays a determinative role in the later development of the trophoblast.

TABLE 4-1. Summary of differences between isolated inner cell mass and trophoblast

Type of tissue	Secretion of blastocoelic fluid	Fusion with fragments of like tissue	Initiation of implantation in recipient uteri
Trophoblast	+	−	+
Inner cell mass	−	+	−

Trophoblastic fragments or vesicles implant in host uteri with a similar frequency to intact blastocysts. They evoke typical decidual swellings in the endometrium, but these lack embryo, amnion, allantois and yolk sac, a result that accords with the presumed origin of such structures from the inner cell mass. Unexpectedly, these implants also show a very conspicuous lack

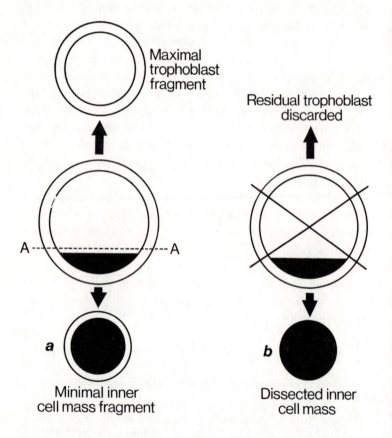

Fig. 4-1. Operations to separate trophoblast and inner cell mass.
a Blastocysts are cut in two along line A---A to obtain pure trophoblast (maximal trophoblast fragment).
b Blastocysts are dissected open to make possible isolation of enclosed inner cell mass tissue (black).

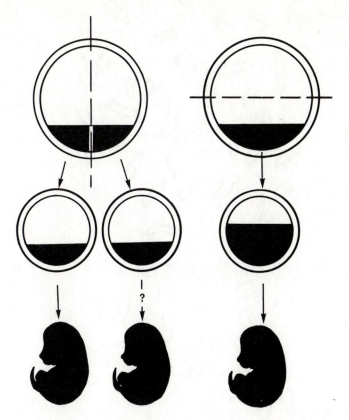

Fig. 4-2. Mouse fetuses and young may be obtained from half blastocysts (*left*) or from blastocyst fragments containing the inner cell
mass (*right*). Whether identical twins can actually be produced from
both halves of a blastocyst, as suggested on the left, has not yet been
established.

of trophoblast derivatives, and contain, at most, a dozen
trophoblastic giant cells. Failure to proliferate cannot be
attributed simply to a deficiency of cells in the transferred
fragments, since partial blastocysts can develop into foetuses or
young (Fig. 4-2), though they may contain fewer trophoblast
cells. Another difference exists between the trophoblastic
fragments that fail to proliferate and intact or partial blastocysts

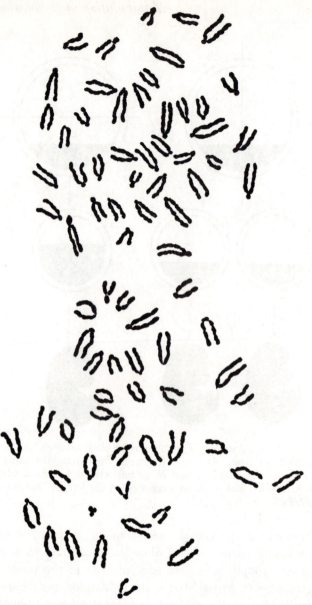

Fig. 4-3. Normal mouse metaphase spread of forty chromosomes (*above*). Normal x CBAHT₆ mouse metaphase showing the distinctive minute T₆ chromosome (*below*).

that do: the former lack not only inner cell mass tissue but also the trophoblast immediately overlying it. The development of 'reconstituted' blastocysts discussed below shows that the presence of inner cell mass tissue is indeed important in this connection.

Development of blastocysts with two inner cell masses

Not only may an inner cell mass be isolated by microsurgery, but it can then be transferred to the cavity of another blastocyst. The latter is thereby induced to carry two inner cell masses. Provided donor and host embryos have distinct cell markers the history of both inner cell masses may be traced after transferring the experimental blastocysts to recipient uteri. Unequal sizes in the smallest pair of chromosomes provides one such marker (the 'T$_6$ translocation', see Fig. 4-3), and another is conferred by the presence of genes for skin, eye and hair pigmentation as distinct from albinism. The procedure for transferring the inner cell mass is shown in Fig. 4-4.

Donor and host inner cell masses often remain separate when the blastocysts are transferred back into the uterus following re-expansion in culture, but they apparently fuse together later because some foetuses and young show both markers, i.e. they are chimaeras. Twins have never been produced as a result of transferring a second inner cell mass into a normal blastocyst. Since the donor inner cell mass can be at least 24 hours older than the host blastocyst, this experiment suggests that the axis of the mouse embryo is not determined until after implantation. This is also probably true in man.

At the beginning of implantation the inner cell mass of the mouse blastocyst is randomly orientated. Later it is invariably directed towards the mesometrial surface of the uterus. What determines this position is a mystery, but it is probably achieved by active migration of the inner cell mass over the inner surface of the trophectoderm (trophoblast). This hypothesis, originally proposed by the late David Kirby and

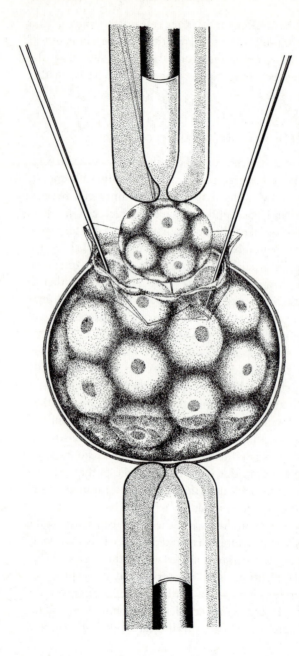

Fig. 4-4. Technique for transferring an inner cell mass to a blastocyst or trophoblastic vesicle. The blastocyst is held by suction on the tip of one smooth glass pipette (*left*) and the donor inner cell mass on another (*right*). Three sharp glass needles are used to make a triangular hole in the trophoblast wall and overlying zona pellucida, clear of the inner cell mass of the host blastocyst. The donor inner cell mass is then introduced into the blastocoele through this hole. Essentially the same procedure is used for transferring cells, except that the right-hand pipette is wider at the tip to accommodate the donor cells within it.

his colleagues in Oxford, would explain how a transplanted inner cell mass can fuse with a host one. The idea is supported by observations of Eric Jenkinson and Ian Wilson of Bangor on blastocysts cultured in small pieces of bovine lens. They noted gross movement of inner cell masses in relation to trophoblast cells that had been marked with oil globules in these blastocysts.

Development of 'reconstituted' blastocysts

An obvious criticism of the preceding experiments is that isolation, culture and manipulation are liable both to alter the properties and impair the capacity for further development of trophoblast and inner cell mass tissue. This is inherently unlikely because even parts of blastocysts can develop normally (Fig. 4-2), as can those from which part of the inner cell mass has been removed by microsuction. The decisive experiment, however, would be to examine the development of blastocysts reconstituted from separated inner cell mass and trophoblast tissue. The procedure is illustrated schematically in Fig. 4-5.

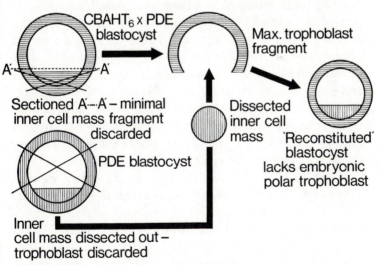

Fig. 4-5. Scheme of experimental 'reconstitution' of mouse blastocysts from trophoblast and inner cell mass.

Manipulation of development

Isolated fragments of trophoblast are cultured until they form vesicles, and an inner cell mass is then inserted into each by the same technique that was used for transferring second inner cell masses to blastocysts (Fig. 4-4).

More than half such 'reconstituted' blastocysts that implant yield fetuses, around which trophoblast development has proceeded normally. Three conclusions can be drawn from this result. Isolation and culture of inner cell mass and trophoblast does not impair their viability. An inner cell mass can develop into a normal fetus when surrounded by a genetically dissimilar envelope of trophoblast. Finally, the proliferation of trophoblast depends on the presence of inner cell mass rather than upon the presence of the region of trophoblast originally overlying it, since the latter tissue is missing from 'reconstituted' blastocysts (see Fig. 4-5).

Injection of cells into the blastocyst

Analysis of early development may be extended yet further by injecting cells rather than tissues into host blastocysts. In preliminary experiments donor blastocysts carrying suitable genetic markers were disaggregated and as many cells as possible (usually two to five) were injected into the blastocoelic cavity of each host blastocyst. Approximately 17 per cent of the fetuses and young that developed from such blastocysts were chimaeric, so some donor cells can obviously colonize host embryos when injected into the blastocoele. By disaggregating isolated trophoblast or inner cell mass tissue, the fate of cells of either type can be studied following injection into the blastocoele.

It was found that trophoblast cells do not colonize embryos, presumably because they are already committed to a different path of differentiation. As one might expect, cells of inner cell mass origin can contribute to the host embryo. Indeed, just one donor cell injected into each blastocyst can lead to extensive colonization in more than 25 per cent of the resulting young. This dramatic result suggests that the number of cells that

118

contribute to the definitive embryo is low in the $3\frac{1}{2}$-day blasto-cyst. It also enables us to follow the descendants of a single cell (known collectively as a clone) *in vivo*, as we shall see later.

Development of the mouse blastocyst – a summary

The following sequence of events is suggested by the preceding experiments and by other observations on the mouse blastocyst. Initially, all the cells of the trophoblast are equi-potential, though clearly differentiated from those of the inner cell mass (Table 4-1). The inner cell mass is largely undifferentiated and its position is variable when the blastocyst first attaches to the uterine epithelium. The inner cell mass soon migrates round the inner surface of the trophoblast until it lies mesometrially (Fig. 4-6a, b and c). Several hours later all cells of the tropho-blast, except those immediately overlying the inner cell mass have become transformed into so-called primary giant cells (Fig. 4-6d). The ectoplacental cone, which gives rise to the beginnings of the embryonic placenta as well as large numbers of secondary giant cells, develops later by proliferation of the trophoblast overlying the inner cell mass (Fig. 4-6e).

The function of the primary giant cells is not known. They have never been seen to divide though they increase considerably in both size and DNA content. Thus, once transformation has taken place, affected cells have forfeited the chance of leaving descendants by mitosis. The fate of trophoblast separated from inner cell mass tissue was, as noted earlier, to form at most a few giant cells. This may be explained most simply by supposing that all the cells of the trophoblast are destined to become giant cells unless they come into intimate contact with inner cell mass. The role of the latter could thus be to prevent transformation or promote division in the adjacent trophoblast cells, which then form the ectoplacenta. This system appears to have the attributes of an inductive process (see Fig. 4-6a to e), of which a classical example is the development of the lens in the verte-brate eye. Alternatively, the inner cell mass might actually

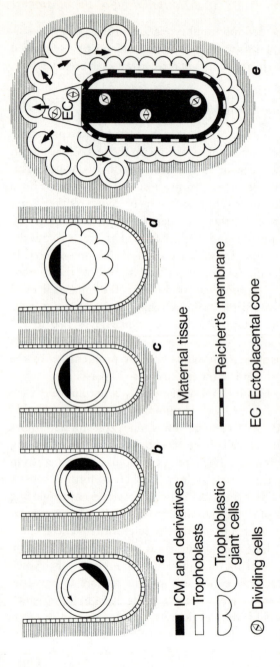

Fig. 4-6. Stages of trophoblast differentiation in the mouse. The blastocyst comes to lie in a uterine crypt with its inner cell mass directed randomly (a). The inner cell mass migrates round the inner surface of the trophoblast until it lies mesometrially (b and c). All cells of trophectoderm except those immediately overlying the inner cell mass are transformed into giant cells (d). The blastocyst then develops into an egg cylinder by downgrowth of the inner cell mass into the blastocoelic cavity, which elongates to accommodate it. The ectoplacental cone forms over the inner cell mass, and gives rise to the definitive trophoblast and secondary giant cells (e). Reichert's membrane is produced by the endoderm and separates this layer from the trophoblastic giant cells.

■ ICM and derivatives   ▥ Maternal tissue

□ Trophoblasts   ▬ ▬ Reichert's membrane

◯ Trophoblastic
giant cells

⊗ Dividing cells   EC Ectoplacental cone

provide the cells from which the ectoplacenta and its derivatives develop.

The existence of some sort of functional dependence of trophoblast on inner cell mass may be advantageous, as it might well ensure harmony of development between trophoblast and embryo.

Blastocyst biopsy – practical applications

The mouse blastocyst is evidently very resilient to microsurgery, and rabbit and sheep blastocysts can also develop normally after the removal of some trophoblast. This raises the possibility of studying the excised tissue to obtain information about the embryo. The type of information that can be gained will depend both on the amount of tissue available and also on the genes that it expresses. Such an approach offers the important advantage over amniocentesis (see Book 5, Chapter 4) that implantation has not yet taken place, and embryos may be selected for transfer according to the results obtained.

The most obvious way that this technique might be applied is in the control of the sex of offspring. Hitherto all attempts to do this have been directed without success towards separating spermatozoa bearing X- and Y-chromosomes. The author, together with Robert Edwards in Cambridge, was able to sex live rabbit blastocysts by a microsurgical technique. Each blastocyst was held by suction on the tip of a glass pipette, and a column of trophoblast was drawn out by suction from the opposite side. The exposed trophoblast was excised with a pair of very fine scissors (Fig. 4-7), and the sample was stained and scored for sex chromatin. The 'sexed' blastocysts were transferred to recipient females after they had recovered in culture. Eighteen developed to term, and the sex of each was found to have been predicted correctly at the blastocyst stage.

This particular procedure cannot be applied to the blastocyst in all mammals – in most species studied sex chromatin does not appear until or after implantation, and there are fewer cells

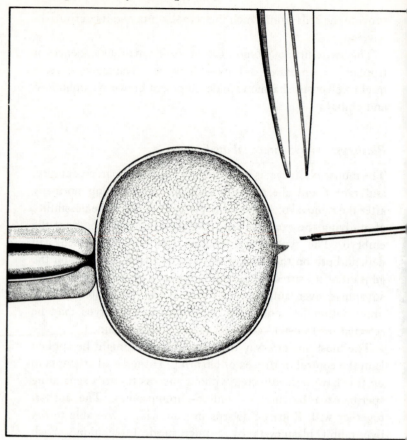

Fig. 4-7. Removal of piece of trophoblast for sexing the living rabbit blastocyst. The blastocyst is held by gentle suction (*left*) while a column of trophoblast is withdrawn from the opposite side. The exposed trophoblast is then excised with a pair of fine scissors. (From R. L. Gardner and R. G. Edwards, *Nature, Lond.* **218,** (1968).)

in whole blastocysts of rat, mouse, hamster, guinea pig and man than in an average biopsy specimen from the much larger rabbit blastocyst. Examination of metaphase chromosomes is an alternative method of sexing, but again only if enough cells can be obtained directly or following culture of biopsy material. Moreover, chromosome techniques are rather capricious, and the development of means for storing blastocysts would be necessary if the excised tissue had to be cultured.

Nonetheless, if developed in conjunction with techniques of superovulation and non-surgical transfer, a procedure based on typing blastocysts could assume practical as well as scientific value. For example, certain serious hereditary diseases in man such as haemophilia and the Duchenne type of muscular dystrophy are due to recessive X-linked genes, and thus afflict sons much more frequently than daughters. Selection of daughters in women who are known to carry the defective gene could avert the birth of affected individuals in perhaps a more acceptable way than the present methods of amniocentesis and abortion. The social and ethical implications of the selection of embryos is beyond the scope of this chapter (see Book 5, Chapter 6).

EXPERIMENTAL CHIMAERAS

Perhaps the most dramatic manipulations of mammalian development are those by which usually two, and possibly more, populations of cells carrying prescribed genetic differences are made to coexist in a single individual from a very early stage. Such chimaeras, as they are called (for further description see Chapter 5), are proving immensely valuable in the investigation of several basic problems of development. They may be produced by fusing together pairs of cleaving embryos (as described in Chapter 1), or, as mentioned earlier, by transferring extra inner cell masses or cells to host blastocysts. The mouse is again the ideal species for these experiments because of its short gestation interval and wealth of known genes.

Chimaeras may arise spontaneously in mammals in several different ways, most frequently by exchange of cells through vascular connections between foetuses (as between dizygotic twins in cattle – discussed in Chapter 2) or by passage of cells in either direction across the placenta. Chimaerism is limited in these instances to one or few tissues, notably blood and trophoblast. A similar degree of chimaerism may be produced experimentally by injecting tissue into neonatal animals whose immunological system is not yet mature, or by applying like measures to adults treated with X-rays or immunosuppressive drugs.

The situation is very different in chimaeras produced experimentally at or before the blastocyst stage, as also in certain natural chimaeras, arising by embryo fusion or by double fertilization after immediate cleavage (see Book 1, Chapter 5). Here all or most tissues may contain a mixture of cells, including the germ line, evidently because the cells come together before the various tissues of the embryo are determined. Nonetheless, not all chimaeric blastocysts give rise to offspring displaying both cell lines; in some cases one or other line is either lost or reduced to an undetectable level during development.

Since the mixing of cells preceded the differentiation of the immune system, chimaeras may show 'intrinsic' tolerance. Provided the constituent genotypes are derived from different inbred strains of mice, grafts from individuals of either parental strains will be retained indefinitely. Unrelated grafts are rejected as vigorously as by any conventional mouse. The most astonishing feature of these chimaeras is, however, that they develop into normal adults which can provide unique information about their developmental history. The number and diversity of genetic differences that can be united in a single embryo is under the control of the investigator, and is in principle almost unlimited.

Differentiation of skeletal muscle in vivo

Skeletal muscle fibres are long spindle-shaped structures containing a central bundle of contractile apparatus and peripheral array of nuclei. These fibres develop in culture by the fusion of several separate myoblasts (precursor cells), each of which contains a single nucleus, rather than from a single myoblast by repeated nuclear division without cytoplasmic

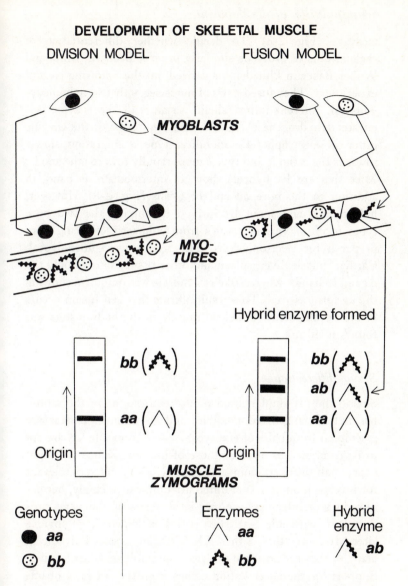

Fig. 4-8. Division (*left*) and fusion models (*right*) to explain the development of skeletal muscle in mouse chimaeras, together with the expected electrophoretic patterns of isocitrate dehydrogenase ('muscle zymograms'). The fusion model was found to be correct. (After B. Mintz and W. W. Baker. *Proc. Natn. Acad. Sci. U.S.A.* **58,** 592 (1967).)

cleavage (Fig. 4-8). To decide whether differentiation of skeletal muscle proceeds similarly *in vivo*, Beatrice Mintz and Wilber Baker in Philadelphia carried out the following elegant experiment. They fused pairs of mouse eggs that were homozygous for different forms (allelic variants) of the enzyme isocitrate dehydrogenase (IDH). The two forms of the enzyme differ in electrophoretic mobility, type *a* migrating slowly towards the cathode and type *b* more rapidly towards the anode. Mice that are F1 hybrids show an intermediate *ab* band, in addition to the pure *aa* and *bb* forms (Fig. 4-8). However, hybrid enzyme cannot be formed by mixing the two pure types *in vitro*. Various tissues from chimaeric offspring were prepared for electrophoresis. The results were unequivocal. Cardiac muscle, liver, spleen and other tissues showed only the *aa* and *bb* bands, whereas skeletal muscle was unique in showing the *ab* form as well. This result affirms that cell fusion occurs in the development of skeletal muscle in the body just as was found in culture.

Sexual differentiation

Sexing the pre-implantation mouse embryo, as in the rabbit, has not so far proved possible, and so when chimaeras are assembled from pairs of embryos, those of opposite sex are apt to be combined as often as those of like sex. Hence we might expect half the overt chimaeras to be XX/XY, and very relevant for investigation of sexual differentiation. Surprisingly, hermaphrodites or intersexes are rare in all series of chimaeras produced so far (Table 4-2). How is this explained? Two obvious alternatives are that, either XX/XY chimaeras mainly die *in utero*, or they are born but pose as respectable males or females. Support for the latter notion comes from the sex ratio of offspring, which shows a significant excess of males in several studies (Table 4-2). In view of the dominant role of the Y-chromosome, Krystof Tarkowski in Warsaw suggested that XX/XY mosaicism may lead to a male phenotype. Examination

Experimental chimaeras

TABLE 4-2. Sex ratio of offspring obtained from chimaeric blastocysts

Authors	Females	Intersexes	Males	Total
Tarkowski (1964)	2	3	11	16
Mystkowska and Tarkowski (1968)	6	1	17	24
McLaren and Bowman (1969)	1	0	13	14
Mintz (1969)	151	4	142	397
Gardner (unpublished)	49	0	50	99

of metaphase chromosomes from bone marrow confirmed that some fertile males were indeed of this type. The story is, however, not quite as simple as this because in Beatrice Mintz's extensive series of chimaeras, she found a normal sex ratio (Table 4-2). Thus, she found XX/XY females as well as males. Why the sex ratio should differ so much between studies is not understood.

Beatrice Mintz further demonstrated that XX/XY male chimaeras may be fertile with histologically normal testes, even though the latter contain a high proportion of XX cells. This raises the interesting question of the origin of the functional germ cells in these animals: are they likewise a mixture of male and female cells? Chimaeras obtained after union of embryos of the same sex often show chimaerism in the germ line as in other tissues, and two genetically distinct types of gametes may be produced by one gonad.

The results of test-breeding fertile XX/XY chimaeras are unambiguous. Their offspring have a normal sex ratio and are all of one type, which invariably corresponds with the gonadal sex of the chimaera. If the gonads are testes, XY cells provide the germ line; if ovaries, the XX germ cells are functional. Functional reversal of the differentiation of germ cells does not occur in the mouse, nor apparently in several other mammals in which autosomal genes can cause testicular development in genetic females (see Chapter 2).

127

Manipulation of development

The fate of germ cells of the inappropriate sex in XX/XY chimaeras is uncertain, and has been studied only in males. Ewa Mystkowska and Krystof Tarkowski could find only XY cells among first meiotic metaphases in testis preparations from adults. However, on examining the testes of late fetuses they found germ cells in meiotic prophase side by side with normal pre-spermatogonia, and it may be significant that precocious meiotic behaviour of this kind is typical of female germ cells at this stage. It is uncertain at present whether these cells are XX germ cells entering meiosis on time, in defiance of the testicular environment around them, or whether they are of both sexes and prompted to behave thus by conditions prevailing locally in an XX/XY testis. What is certain is that such 'oocytes' disappear soon after birth.

Clonal aspects of development

When equal numbers of two populations of cells are present in embryos before particular tissues or organs differentiate, analysis of the cellular composition or pattern of these structures in adults can provide an estimate of the number of precursor cells from which they developed. Thus if a tissue developed by division of only one precursor cell it should be composed exclusively of cell type A in half the offspring, and of type B in the remainder. A tissue developing from two precursor cells would be of type A in a quarter, type B in a quarter, and a mixture of A and B in a half. As the number of precursor cells increases, the frequency of adults showing only type A or B cells in the tissue will decrease as indicated in Table 4-3. In practice the two types of cell are unlikely to divide at the same rate so that the composition of the adult tissue will differ from that of the embryonic rudiment. Nevertheless, application of this principle to chimaeras – and to 'mosaics' arising from X-inactivation (see next section) – leads to the conclusion that many tissues and organs originate by division of relatively few foundation cells. Each precursor cell together with its progeny is called a clone.

TABLE 4-3. Relationship between precursor cell number and adult composition of tissues in chimaeras or mosaics

No. of cells	Approximate frequency of individuals with cells all one type (A or B)
2	50%
3	25%
4	12.5%
5	6.25%
6	3.1%
8	1%
10	0.2%
20	2×10^{-6}%

Clonal development is illustrated most clearly when the constituent cells of chimaeras or mosaics carry different coat colour genes such as albinism versus pigmentation. Patterns are then clearly visible on the surface of the living animal. Pepper-and-salt markings due to complete intermixing of albino and pigmented hairs are never seen. Instead, more or less discrete patches of one or other colour is the rule. These tend to be arranged transversely and are often asymmetrical thereby revealing a discontinuity along the dorsal midline (Fig. 4-9). Such patches arise through clonal proliferation of pigment precursor cells that have migrated laterally from the neural crest in the embryo. The patches will obviously vary in size depending on whether adjacent clones are of the same or opposite type.

Investigation of X-inactivation

In female mammals one X-chromosome is inactivated early in development and appears in the nucleus as the sex chromatin

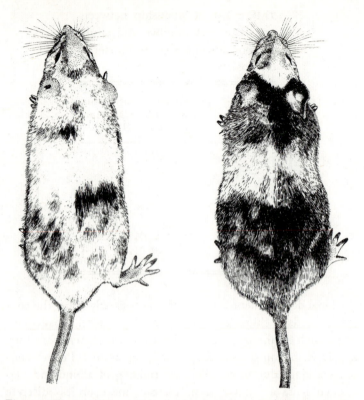

Fig. 4-9. Mouse chimaeras obtained by transferring a single inner cell mass cell from a pigmented to an albino blastocyst (*left*), or by transferring an entire albino inner cell mass to a pigmented blastocyst (*right*). Pigmented and albino areas are organized into discrete patches on either side of the midline.

or Barr body. The production of mouse chimaeras by injecting one cell into each host blastocyst provides a means of determining the time of X-inactivation in this species – hitherto this could only be approximately inferred by noting when sex chromatin first appeared. The principle of the test is so to arrange mating that donor embryos, if female, will have two distinct coat colour genes controlled by their two X-chromosomes. A single donor cell is then injected into each of a series of host blastocysts that carry a third unrelated colour. If the female donor cell has undergone inactivation, the resulting chimaera will show one of the two donor colours together with the host colour. Otherwise a three-coloured mouse will be expected. Preliminary results of experiments performed in collaboration with Mary Lyon of Harwell indicate that X-inactivation has not taken place in the 3½-day mouse blastocyst, though it probably does so a few cell divisions later. Also, astonishingly, over one-quarter of the offspring that developed from blastocysts receiving a single donor cell were overt chimaeras.

TRANSPLANTATION OF NUCLEI

The technique of injecting isolated nuclei into enucleated eggs was devised originally for the large amphibian egg by Robert Briggs and Thomas King working in Philadelphia. Two main conclusions emerge from the amphibian studies undertaken in several laboratories: (1) Nuclei from highly specialized late embryonic or adult cells can support normal development of host eggs. Hence cellular differentiation does not necessarily involve irreversible change in or loss of genetic material. (2) Early development in amphibia is largely controlled by factors located in the egg cytoplasm.

Transplanting nuclei to eggs is obviously a very useful means of analysing nuclear changes and relations with the cytoplasm during development. However, its application to mammals poses formidable technical difficulties, not least of which is inherent in the small size of their eggs. Nonetheless, the first steps have been taken by Christopher Graham of Oxford University using mouse eggs. He exploited the fact, first discovered by Henry Harris and John Watkins, also at Oxford, that killed 'Sendai' virus retains the ability to fuse cells together. The donor cells were thus coated with virus particles and then encouraged to fuse with fertilized or unfertilized eggs previously denuded of the zona pellucida. The host eggs were not enucleated in preliminary experiments.

Manipulation of development

Donor cells fused with 1-cell eggs first undergo nuclear swelling as in amphibia, and the donor nucleus then divides once in synchrony with the host nucleus. Development stops at the 2-cell stage – further progress could indeed prove possible but, with present methods, even normal mouse eggs cultured from shortly after fertilization typically arrest at the 2-cell stage (see Chapter 1).

Thus, although the initial behaviour of transplanted nuclei is encouraging, several obstacles must be surmounted before mammalian offspring can be obtained from eggs containing transplanted nuclei.

The past decade has witnessed an explosion of interest in the early mammalian embryo as an object for experimental investigation. The remarkable resilience of the cleaving egg and blastocyst may be exploited in several different ways to gain information on development and exert some degree of control. Experimental chimaeras are probably the most valuable products of these studies so far. Their potential for the analysis of development as well as for research on cancer and problems of immunology is only just beginning to be realized, and is likely to be considerable. Further rewards are just around the corner. The repair of genetically faulty blastocysts by the inoculation of cells from normal ones may soon be possible. Nuclear transplantation could have tremendous impact by allowing the immediate and precise selection of genotype. Future prospects in this field look bright indeed.

SUGGESTED FURTHER READING

Choosing sex before birth. R. G. Edwards and R. L. Gardner. *New Scientist* **38**, 218 (1968).
Developmental implications of multiple tissue studies in glucose-6-phosphate dehydrogen? ?-deficient heterozygotes. E. Gandini, S. M. Gartler, G. Angioni, N. Argiolas and G. Dell'Acqua. *Proceedings of the National Academy of Sciences, U.S.A.* **61**, 945 (1968).
Mouse chimaeras obtained by the injection of cells into the blastocyst. R. L. Gardner. *Nature, London,* **220**, 596 (1968).

Suggested further reading

Manipulations on the blastocyst. R. L. Gardner. In *Advances in the Biosciences* 6. Schering Symposium on Intrinsic and Extrinsic Factors in Early Mammalian Development. Venice, 1970. Ed. G. Raspé. Vieweg, Pergamon Press (1971).

Transplanted nuclei and cell differentiation. J. B. Gurdon. *Scientific American* **219**, 24 (1968).

Observations on CBA-p/CBA-T6T6 mouse chimaeras. E. T. Mystkowska and A. K. Tarkowski. *Journal of Embryology and Experimental Morphology* **20**, 33 (1968).

Gene action in the X-chromosome of the mouse (*Mus musculus* L.). M. F. Lyon. *Nature, London* **190**, 372 (1961).

Developmental mechanisms found in allophenic mice with sex chromosomal and pigmentary mosaicism. B. Mintz. In *Birth Defects*. Original Article Series **5**, 11 (1969).

Normal mammalian muscle differentiation and the gene control of isocitrate dehydrogenase synthesis. B. Mintz and W. W. Baker. *Proceedings of the National Academy of Sciences, U.S.A.* **58**, 592 (1967).

Germ cells in natural and experimental chimaeras in mammals. A. K. Tarkowski. *Philosophical Transactions of the Royal Society of London* B. **259**, 107 (1970).

5 Pregnancy losses and birth defects
C. R. Austin

Prenatal development is particularly hazardous: in man the chances of death are greater before birth than at any other time in life except for extreme old age, and in other animals too the magnitude of prenatal loss is often remarkable even under natural conditions. Exactly how many early embryos die is difficult to say because they simply fail to implant or are reabsorbed, causing no outward sign and leaving few internal traces. An informed guess puts the human mortality at this time at about 30 per cent; losses later in pregnancy appear as abortions, and these, taken with still-births, account for some 20 per cent of established pregnancies. So about half of all fertilized human eggs fail to produce embryos that survive, even in the absence of birth control measures. More systematic investigations are clearly possible in certain other animals, and the data show that prenatal losses ranging between 15 and 60 per cent occur in cattle, sheep and pigs, as well as in wild forms such as stoats, rats, squirrels and rabbits.

The dangers to life before birth are not only those of premature death, but also of maldevelopment that can lead to many kinds of birth defects, from insignificant 'impediments' to crippling physical, mental or functional disorders. This is a matter of special importance in man. About 5 per cent of children born alive exhibit recognizable abnormalities (termed congenital to denote their existence from birth). Many of these children soon die of their defects, while others will drag on into later years maimed and handicapped. In contrast to the large death rate before birth, the 5 per cent figure may seem small, but is in fact a far greater tragedy both from the personal and national viewpoint. In the United Kingdom with a birth rate of approximately 1 million per year, 5 per cent means that

50 000 abnormal children are born each year, leading to a floating population of nearly half a million afflicted people – a sad burden on relatives, friends, society and themselves.

Birth defects are not a human prerogative, and a wide variety has been recorded in the newborn of other animals. But their occurrence in animals is less frequent than in man and afflicted individuals soon fall victim to predators in Nature or are hastily disposed of under conditions of domestication; they rarely constitute serious economic loss. Consequently the main emphasis in this chapter is on developmental irregularities in man.

There are many reasons for disturbance of normal development. Those that we seem to understand best can be placed in four groups: maternal limitations, chromosomal errors, inherited defects, and damaging agents. The classification is for convenience and should not be taken to imply the existence of distinct and separate categories. Most disorders are brought about by several different influences, acting together or in

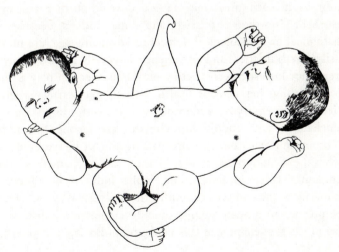

Fig. 5-1. Conjoined twins. (From F. L. Potter, *Pathology of the Fetus and the Infant*, 2nd edition. Year Book Medical Publishers, Chicago (1961). This figure courtesy of Dr Hans G. Schlumberger and J. Elmer Gotwals.)

succession. Often there is an inherited predisposition, but what precipitates the condition may be quite unknown. Identical twins, for example, have a tendency to run in families, but why the twinning mechanism should sometimes malfunction and result in conjoined twins (Fig. 5-1) we cannot say. Occasionally the secondary cause can be recognized as a dietary deficiency or a drug of some kind. Often, the very existence of a disability reduces resistance to infection, and so the case is worsened by the ravages of bacteria or viruses.

MATERNAL LIMITATIONS

Under this heading we are concerned with the kind of environment that the mother provides for her future offspring throughout pregnancy; during most of this time the uterus makes up the new individual's whole world and exerts influences on development. Often in polytocous animals, which have litters, such as rabbits and pigs, the crowding is such that several embryos in each pregnancy fail to develop further and are resorbed. Crowding is relative, and some mothers, because of anatomical or physiological defects or advancing age, are more easily overburdened and experience larger or more frequent pregnancy losses. Stresses acting on the mother can play a role by disturbing her general health without necessarily directly affecting the embryos; undernutrition or overfeeding, vitamin deficiencies, and unduly hot climates are among the more common indirect causes of prenatal death (or, in the event of survival, of abnormal young). Other causes, named in the remaining three groups, entail faults induced or inherent primarily in the embryos or fetuses themselves. (Ways in which the uterus can influence embryonic development are discussed also in the first chapter of this book and in Book 4, Chapter 5.)

CHROMOSOMAL ERRORS

Since chromosomes are the mediators of genetic influence, even small disturbances in their number and form interfere with normal development. Major chromosomal errors arise, as indicated in the last chapter of Book 1, in the maturation of the egg and the course of fertilization. The frequency of errors can be judged from the fact that in man, nearly 42 per cent of aborted fetuses and about 1 in every 200 newborn children show departures from normal in chromosomal state.

Chromosomes are most conveniently studied in standard arrays or karyotypes, which can be prepared from cells of many different tissues (see the second chapter of this book). The cells used most often are the white cells of the blood, the polymorphonuclear leucocytes, as these can easily be grown in culture; another source, of special significance in the present context, is the amniotic fluid in which cells from the developing fetus abound. This fluid can be collected by the surgeon using a relatively simple procedure known as amniocentesis, as early as about the 4th month of pregnancy. If chromosomal errors are found, the law in some countries permits pregnancy to be terminated at this stage, thus saving the birth of a deformed child.

The normal human complement is forty-six chromosomes, the total being made up of twenty-two pairs of autosomes and a pair of sex chromosomes, conveniently expressed as 46XX for a woman or 46XY for a man. The normal chromosomal status can be disturbed in a number of different ways, the most important being the following:

Trisomy and monosomy

These terms imply that there is either one additional member to what is normally a pair of chromosomes, or only a single chromosome instead of a pair. The error arises through a failure of separation, non-disjunction, of chromosomes at one or other of the two meiotic divisions, in oogenesis or spermatogenesis. If egg or spermatozoon carries a chromosome group with an additional member, the embryo will display the state of trisomy. Alternatively, if egg or spermatozoon lacks a particular chromosome, the embryo will exhibit monosomy.

When the disturbance is of autosomal chromosomes the consequences are severe: in man trisomic embryos generally die before birth, and monosomic ones apparently always do. The commonest condition compatible with survival is trisomy-21, namely trisomy involving the 21st pair of chromosomes which gives rise to the condition known as mongolism or

Fig. 5-2. A group of mongoloid boys. (From M. Engler. *Mongolism* (*Peristatic amentia*). Wright, Bristol (1949).)

Down's syndrome. Such children show a number of physical and functional disabilities as well as being seriously retarded mentally, hence the alternative name mongoloid idiocy (Figs. 5-2 and 5-3). About 1 in 700 births is a mongol and the frequency shows a curious tendency to increase with maternal age, so that among babies born to women of 45 years or more the frequency is around 1 in every 50 (this point is taken up again in Book 4,

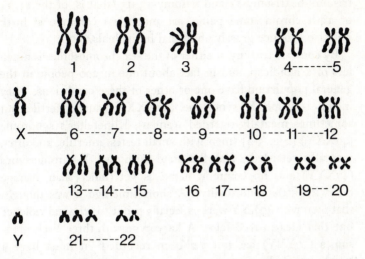

Fig. 5-3. The karyotype of trisomy-21, Down's syndrome or mongolism. (From P. E. Polani. In *Birth Defects*. Ed. M. Fishbein. Lippincott, Philadelphia (1963). By permission of the National Foundation, New York.)

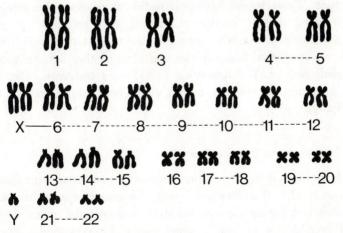

Fig. 5-4. The karyotype of Klinefelter's syndrome (47XXY). (Same source as Fig. 5-3. By permission of the National Foundation, New York.)

139

Chapter 5). The other autosomal trisomies that survive – trisomy-6, trisomy-17 and trisomy-13–15 (that is, of the 13, 14 or 15th chromosome pair) – are much less common at birth but also produce grossly abnormal individuals.

By contrast trisomy of either of the sex chromosomes is much less of a handicap, and in fact about one in 200 people in the general population have one or other of these conditions. They include: Triple-X syndrome (47XXX) (usually fertile but sometimes mentally-retarded women), Klinefelter's syndrome (47XXY) (Fig. 5-4) (men with small testes, infertile, and often mentally retarded), and the state with two Y-chromosomes (47XYY) and no common name as yet (taller than average but with little abnormality). A short time ago it was thought that men with 47XYY were especially prone to criminal violence but this idea proved false. A karyotype with three Y-chromosomes (47YYY) has not yet been recorded – it must have a highly lethal effect.

The only sex-chromosome monosomy that is consistent with survival is one in which there is a single X-chromosome; this is Turner's syndrome (45XO). It affects about one woman in 1000. These people have only rudimentary ovaries containing no germ cells at the time of puberty; they are physically undeveloped as females, unusually short in stature, and often have congenital heart disease and certain other defects. Cases with only a Y-chromosome (45OY) are unknown – the X-chromosome evidently carries vital genes and so cannot be spared.

Inversions and translocations

Sometimes in the formation of egg or spermatozoon, for reasons not at all well understood, a part of one chromosome becomes detached during meiosis, and then joins on again upside down (inversion) or to another chromosome (translocation). Inversions nearly always kill the embryo at an early stage, but the effect of a translocation can be rather similar to that of having a whole extra chromosome, namely to that of a trisomy. Only two viable translocations are known, and these are identified as 15/21 and 13/15 (part of chromosome 15 added to chromosome 21, or part of 13 added to 15). People with 15/21 translocation show a condition that resembles mongolism quite closely, and cases with the 13/15 translocation are rather like those with trisomy-13–15.

Triploidy

In this condition there are trios of each type of chromosome instead of pairs (69XXX, 69XXY or 69XYY). The two most likely ways in which these things can come about are through polyspermy and polygyny (see Book 1, Chapter 5). In polyspermy two spermatozoa take part in fertilization, and in polygyny the first or the second polar body fails to be formed; both processes cause the egg to have an extra chromosome set. Triploid embryos appear to be able to develop in a perfectly normal way to begin with, but the embryo sickens and dies about halfway through pregnancy. There are records of the birth of reputedly triploid human beings; later, however, most were found to have also some diploid cells, and so were not pure triploids but mosaics – just how this state of affairs might arise will be discussed shortly.

Haploidy

This means that the embryo has only half the normal number of chromosomes; the situation usually arises through parthenogenesis or gynogenesis, both of which were described in Book 1, Chapter 5. Briefly, in parthenogenesis the egg enters upon development apparently spontaneously and without the help of the spermatozoon; in gynogenesis the egg is stimulated to develop by a spermatozoon that plays no further part in fertilization. In mammals haploid development is of very short range and has not been reported much beyond mid-pregnancy.

Pregnancy losses and birth defects

Mosaicism and chimaerism

These conditions are probably more common in man than was originally thought, and the terms refer not only to chromosome differences in parts of various tissues or organs but also to differences due to different genes or to gene mutations (we shall have more to say about mutation in the next section). Mosaicism can arise through errors involving mitosis or the occurrence of mutations during the course of embryonic development (so that some of the embryo's cells are different from others). Chimaerism can result from fertilization of each half of an egg in which the second maturation spindle has moved into the centre and caused precocious cleavage of the whole cell; or from fusion of two separate embryos at some stage of cleavage; or from the migration of cells between two fetuses (as in freemartinism – see Chapter 2). The possibilities of embryo fusion taking place, even under normal circumstances, are clearly suggested by the experimental work described in the previous chapter. (The difference between a mosaic and a chimaera is essentially that the first contains the products of one zygotic line, and the second the products of two zygotic lines.)

By whatever means they arise, mosaic and chimaeric individuals can often survive birth and reach maturity – there are now numerous clearly established instances in adolescent and adult human beings. Careful analysis has shown that they are of genetically heterogenous constitution, the mixtures including different kinds of pigmentation in the skin and the iris of the eye, red blood cells belonging to different blood groups, and varying degrees of development of combined ovary and testis tissue. Most of these cases were discovered accidentally, and it is entirely possible that many more people have cells or tissues of different genetic types in their bodies, because mutations or mitotic errors can take place at any time of life. Individuals with these conditions may seem quite normal, for many combinations of cells can manage a peaceful coexistence.

In addition, if there are defective cells in an embryo they can be compensated for to some extent by normal cells, and this is thought to underlie the survival of the largely triploid human children that have been born – they have sufficient diploid cells for the essential jobs that triploid cells may fail to do. They must have had a complex origin involving polyspermy or polygyny, together with double fertilization or embryo fusion.

INHERITED DEFECTS

By and large, inherited defects are due to gene mutations, alterations that nearly always produce disadvantageous characters as compared with the normal gene product. Mutations can occur at any time and may arise for no known reason ('spontaneous' mutations) or be induced by X-rays or other kinds of radiation, or by certain alkylating agents or teratogenic chemicals, such as nitrogen mustards. Mutations are of great importance if they take place in cells of the male or female germ cell line of the adult for they can lead to the death of the embryo arising from the union of affected germ cells, or, if the embryo survives, their influence can recur through succeeding generations. On the other hand, if mutations occur in the blastomeres of early embryos or the cells of later ones, they will affect only that embryo or parts of it.

Hereditary (genetic) defects can take many forms and affect structure or function, or usually, both. Some are due to single gene mutations, and others to more complex situations in which several genes are involved, and outside influences, such as drugs, radiations and disease processes, may also play a role. In man, the well-recognized hereditary conditions vary all the way from small faults like colour blindness, through relatively minor deformities such as harelip (Fig. 5-5), cleft palate, clubfoot (Fig. 5-6), and 'hole in the heart', to more serious defects such as congenital absence of hands, feet or arms, or the strange, remote mental state of autistic children, and major catastrophes such as fibrocystic disease which maims and kills

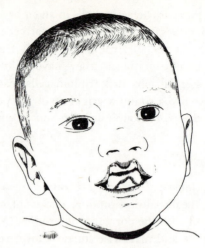

Fig. 5-5. Double harelip. (Same source as Fig. 5-1. © 1961, Year Book Medical Publishers, Inc. Used by permission.)

in childhood (Fig. 5-7), or phenylketonuria which severely and permanently retards growth and mental development. Minor anatomical deformities can be partly or completely corrected by surgery, but we can offer little compensation to those lacking extremities or limbs, other than inadequate artificial substitutes, and we can at the moment offer only limited palliative treatment in conditions such as fibrocystic disease.

On the other hand phenylketonuria, which represents one of a group of anomalies known as inborn errors of metabolism, has a mechanism that can be explained and can be treated with a measure of success. This condition is due to a fault in the body's machinery for dealing with protein foods, in consequence of which one of the component amino acids (phenylalanine) and certain derivatives accumulate to toxic levels and are directly responsible for the symptoms shown. The disease can now be recognized soon after birth, and by a special dietary regime the symptoms can be abated and the course of the disease blocked, so that there are prospects of a more or less normal life ahead. The inborn errors of metabolism are perhaps some of the more

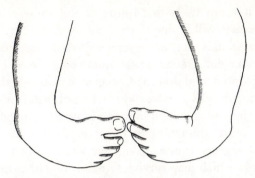

Fig. 5-6. Clubfoot. (Same source as Fig. 5-1. © 1961, Year Book Medical Publishers, Inc. Used by permission.)

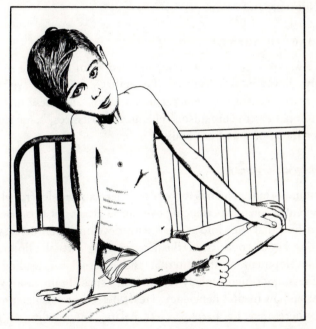

Fig. 5-7. Fibrocystic disease in a 14-year-old boy who had only a few more weeks to live. (Drawn from a photograph in 'Fibrocystic disease' by S. W. Royce. In *Birth Defects*. Ed. M. Fishbein. Lippincott, Philadelphia (1963). By permission of the National Foundation, New York.)

hopeful of the serious developmental problems at the present time, for it seems likely that future careful study of the bio-chemical events will in due course lead to effective treatments.

In other animals there are several well-recognized conditions due to mutation. Dexter cattle quite often have curiously short-legged and round-headed progeny – these are the 'bull-dog' or achondroplastic calves which die just before or at birth. In cattle, pigs and dogs, and in turkeys, there is a mutation that causes abnormal shortening of the spine so that the body of the animal appears 'concertina'd'; this too is a lethal state, ex-cept in dogs which may survive to maturity. Other mutations produce complete absence of limbs in pigs, extra digits in cats, hardening and thickening of the skin in cattle, lack of feathers in chickens, and so on.

DAMAGING AGENTS

Here we are concerned with factors that seem to exert a selec-tively detrimental effect on the embryo or fetus, without necessarily causing the mother much harm, and these include certain pathogenic organisms, chemical substances, X-rays and other radiations, and antibodies.

Pathogenic organisms

Most important in man is the virus of German measles (rubella). Pregnant women experiencing even mild attacks of this disease often find that the child is born with developmental defects of various kinds – opacity of the lens causing virtual blindness, maldevelopment of the internal and external ear leading to deafness, and microcephaly (abnormally small brain case) with its attendant mental deficiency. Heart defects are also common and there may be harelip, cleft palate, abnormality of the intestinal tract, and spina bifida (exposed spinal cord and nerves) (Fig. 5-8). In severely affected cases death may come early. The effects of the disease vary according to the stage of pregnancy when the mother became infected – if in the first couple of weeks,

146

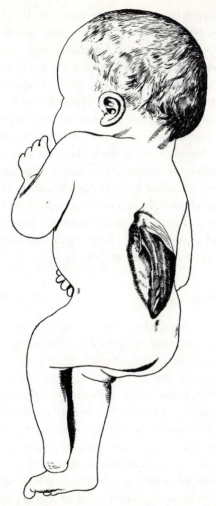

Fig. 5-8. Spina bifida. (Same source as Fig. 5-1. © 1961, Year Book Medical Publishers, Inc. Used by permission.)

or not until the last 2 or 3 months, there may be little or no discernible damage, but if during the 3rd to 12th weeks (the period of embryonic organogenesis, when the principal organs are being established) the consequences can be most serious. From the nature of the deformities shown by affected children,

147

it is clear that there is a limit to what can be expected in the way of remedial treatment, and the only satisfactory method of prevention seems to be to ensure that girls experience German measles before they reach child-bearing age.

Some other viruses have also been suspected of causing birth defects, but there is little evidence for this idea, except in the case of the rather uncommon cytomegalovirus which apparently can damage nerves and brain, leading to mental retardation. Among farm animals there are several viruses that are responsible for reproductive failure, apparently through the induction of lethal embryonic anomalies, but these have not yet been adequately studied. Bacterial conditions may also cause prenatal losses in farm animals, notably brucellosis, vibriosis and leptospirosis; these organisms probably account for about 30 per cent of abortions.

Chemical substances

There are very many chemical substances that can cause birth defects, and they are collectively referred to as teratogens (monster producers); they include mutagenic agents such as nitrogen mustard and trypan blue, metabolic inhibitors such as aminopterin, some hormones such as cortisol, and even some vitamins, especially vitamin A, when given in excess. All these agents, if administered at the appropriate time in pregnancy, can produce gross anomalies such as hydrocephalus (greatly enlarged head due to fluid pressure inside), spina bifida and limb deformities (including ectromelia, where whole limbs are missing; or phocomelia, where arms or legs are lacking and the hands or feet are attached to the trunk). These facts are increasingly becoming a matter for concern because of the growing extent of chemical pollution from industry and agriculture. A rather striking example was reported in Japan a few years ago, under the name of Minimata's disease, the main feature of which was cerebral palsy (paralysis due to damage to the brain) in newborn babies. It was found to be caused by mercury poisoning, the metal being in the effluent from a fertilizer factory discharged into a bay in which many people were accustomed to catch fish. Consumption of the fish by the pregnant women completed the chain.

At this point we should mention the 'spastics' – children who suffer in various degrees from a rigid immobility of muscles (spastic paralysis), usually most evident in the limbs. Spastic paralysis is a product of cerebral palsy and can arise in several ways in addition to mercury poisoning, the common factor being that the damaging agent exerted its effect late in pregnancy, or even soon after birth. Some causes are: bacterial infections, oxygen deficiency as in 'blue' babies (due to heart defect or when the birth process is unduly prolonged), anaemia and jaundice as in Rhesus disease, and low blood sugar levels (hypoglycaemia) such as occur in diabetes.

Various drugs have also been suspected of responsibility for birth defects; in many instances they probably are, but there is a great range in individual susceptibility and here too the stage of pregnancy when the drug is taken is of critical importance. Among the most likely culprits are quinine taken for the relief of malaria, aminopterin administered in attempted abortions, Busulphan (myleran) employed as treatment for leukaemia, and Chlorambucil given to alleviate Hodgkin's disease.

The most dramatically outstanding example of a harmful drug was Thalidomide, which was taken as a tranquillizer and to reduce vomiting in pregnancy. It was highly effective for these purposes, but there is now no doubt that it was also highly effective in producing birth defects – during the period that it was marketed (1958–61) it was responsible for several thousand cases of serious developmental anomaly in West Germany alone, where it was originally manufactured. The most critical period of pregnancy was between the 4th and 7th weeks (during organogenesis), and so the effect was mainly shown in such birth defects as ectromelia, phocomelia, maldevelopment of ears, and numerous internal disturbances. The tragic experience with Thalidomide has greatly sharpened awareness of the need for adequate testing of drugs before they are released for general use, but this is not a simple matter, because many women who took Thalidomide during the critical period of pregnancy had normal children, reflecting large differences in susceptibility, and similar birth defects have been seen, admittedly very rarely, in children born to women who had not taken the drug (see Fig. 5-9).

A curious condition in sheep has been found due to the consumption of the plant *Veratrum californicum* (which contains the potent alkaloid veratrine) – pregnancy is much longer than normal and the young show major deformities of the skull and virtual absence of the pituitary. Much the same set of

Fig. 5-9. A child with phocomelia in all limbs. Drawn from a reproduction of a painting by Goya (1746–1828). (From R. W. Smithells. In *Advances in Teratology*. Ed. D. H. M. Woollam. Logos-Academic, London (1966). By permission of Cabinet des Dessins, Musee du Louvre.)

circumstances is seen in a genetic anomaly of some Jersey and Guernsey calves, which are born with numerous abnormalities including a defective pituitary or total lack of the gland, and here too there is very prolonged pregnancy. Investigations indicate that in both conditions the extention of pregnancy is due to the lack of a functional fetal pituitary, which normally plays an important role in initiating the birth process (further details on this topic are given in Chapter 3 of this book).

X-rays and other radiations

It is not common these days for women to be X-rayed during pregnancy but if it is done great care must be taken to shield the child. X-rays are potent agents for inducing developmental errors. Effects of irradiation are of two kinds, hereditary and direct. Hereditary effects may be due to changes in either chromosomes or single genes; in both instances, the immediate child will not suffer as much as future generations, owing to changes induced in the child's germ cells. Direct effects of irradiation are most clearly seen – like the influence of other damaging agents – if applied during the phase of organogenesis, the embryo then being much more susceptible than the mother. Irradiation at this time can cause a range of gross malformations, with relatively small risk of killing the embryo. By contrast, irradiation in the first 2 weeks of pregnancy, before the embryo is implanted in the uterus, may kill the embryo, but if it does not it is unlikely to have any permanent effect; and irradiation late in pregnancy may cause damage to certain tissues but not to whole organs. Probably the most extensive series of radiation effects in new-born children was seen after the atomic bomb explosions at Hiroshima and Nagasaki. Women who experienced those events during the period of their pregnancy gave birth in many cases to children showing serious defects, the most common being microcephaly (small brain case) with mental deficiency.

Pregnancy losses and birth defects

Antibodies

The immunological relations between mother and fetus have stirred much interest in recent years, and these matters are discussed in other chapters of this book and especially in Book 4, Chapter 4. One of the most important abnormalities affecting this system in man is Rhesus disease, haemolytic disease or erythroblastosis. Haemolytic disease, involving a very similar mechanism to that in man, is known to occur in the horse, though the damaging antibodies are passed to the newborn after birth, in the colostrum, the first milk secreted. Commonly, affected foals show a rapidly developing anaemia and jaundice, and die within a few days of birth. (The involvement of the milk in the transfer of maternal antibodies reflects the greater complexity of the placental barrier which in the horse is not permeable to antibodies.) Conditions closely akin to haemolytic disease have been produced experimentally in the pig, rabbit and dog, but natural occurrence is uncertain. Attempts to induce a similar state in sheep and cows have not been successful.

One's first reaction to figures showing the magnitude of prenatal death is perhaps mystification and sorrow that so many young people and animals fail to see the light of day. To an extent this is a well-founded point of view, for many of the losses are indeed due to environmental factors and involve healthy embryos. But accumulating evidence suggests that the prenatal elimination is in the main an important and valuable provision of Nature. The fact is becoming increasingly clear that a large proportion of resorbed or aborted embryos and fetuses are abnormal, and that their summary disposal is in the best interests of the race. Probably this represents the main way in which disadvantageous features arising from gene mutation are prevented from becoming incorporated into the overall hereditary pattern. Indeed if we must grieve over pregnancy losses it should be rather because so many products of anomalous development still succeed in evading this act of natural selection.

Suggested further reading

Human society is heavily burdened by the horde of unfortunate, pitiful, useless defectives that it is constrained to maintain. Clearly there is a basic social and humanitarian duty to reduce this burden – by therapy where possible, by prevention where possible, but if necessary also by artificially supplementing the natural process of elimination.

SUGGESTED FURTHER READING

Developmental Anatomy, 7th edition. Ed. L. B. Arey. Philadelphia (1965).
Comparative Aspects of Reproductive Failure. Ed. K. Benirschke. New York; Springer-Verlag (1967).
Genetics for the Clinician. C. A. Clarke. Philadelphia; David (1964).
Birth Defects. Ed. Morris Fishbein. Philadelphia; Lippincott (1963).
Pathology of the Foetus and the Infant, 2nd edition. F. L. Potter. Chicago; Year Book Medical Publishers (1961).
Genetics and disease. Clyde Stormont. *Advances in Veterinary Science*, **4,** 137 (1958).
Advances in Teratology (annual volumes, 1966–). Ed. D. H. M. Woollam. London; Logos.

Index

Index

fetus, 72–109
  adrenal glands of, 93, 94, 98, 103, 105
  breathing movements of, 87, 96
  carbohydrate and fat reserves of, 99–101
  cardio-vascular functions of, 81, 82
  chemical substances affecting, 148–51
  damaging agents affecting, 146–53
  defaecation, 83
  effects of antibodies in, 151, 152
  endocrine functions of, 87–94
  gastro-intestinal functions of, 82–4
  glucose in, 73, 74
  glucose utilization by, 74–6
  gonads of, 92, 93
  growth, 73–80
  hyothalamus of, 89, 90, 103
  immunological functions of, 94
  influence on mother, 88, 89
  lung function of, 97–9
  maternal influences on, 89
  movements of, 95, 96
  neurological functions of, 94–7
  pancreas of, 91, 92
  pituitary of, 90, 103, 105, 150, 151
  renal functions of, 85–86
  respiration, 77–9
  sleep, 87, 96, 97
  tactile response, 95
  taste, 83, 84
  temperature, 95
  thyroid of, 90, 91
  urine, 85, 86
fibrinoid, 40
follicle cells, 10
foramen ovale, 81, 82
freemartin, 19, 53–5, 58

gene expression, 14, 15
gene, lethal t^{12}, 15
genetic sex, 44–50
genital ridge, 49
germ cells, 50, 51, 128
german measles, 146–8
gestation, time sequences of, 22
G1 period of mitosis, 3
glucose-6-phosphate dehydrogenase (G-6-PD), 13, 15
glycogen, 11
gonadal inductor substance, 52, 53

gonadal sex, 51–6
gynandromorph, 57
gynogenesis, 141

haemolytic disease, 152
haploidy, 141
hermaphrodite, 43, 53, 54, 126
heterochromatin, 45
heterogametic sex, 44
histotrophe, 25
homogametic sex, 44
homosexuality, 69
hormonal sex, 56–9
human chorionic somatomammotro-phin (HCS), 74
hydrocephalus, 148
hypothalamic sex, 65, 66

immunological enhancement, 41
implantation, 24, 25
  carbon dioxide in, 10
  hormonal control of, 29, 31, 33, 34
  immunological aspects of, 39–41
  induction of, 111
  in ectopic sites, 36–39
inactivation of X-chromosome, 45
inherited defects, 143–6
inner cell mass, 3, 4, 7, 9
  differentiation of, 110–13, 115–21
intersex, 43, 126, 127
interspecific hybrids, 15
inversions of chromosomes, 140, 141

karyotype, 43
kidneys of fetus (see fetal renal functions)
Klinefelter's syndrome, 46, 49–51, 139, 140

lactate dehydrogenase (LDH), 13
lactation, initiation of, 101
legal sex, 69, 70
libido, 67, 68
light, reaching fetus, 95
lordosis, 67
Lyon hypothesis, 45–7, 129–31

male-determining genes, 50
manipulation of development, 110–32
maternal limitations to pregnancy, 136

Index